Unmanned Aerial Vehicle: Applications in Agriculture and Environment

Ram Avtar • Teiji Watanabe
Editors

Unmanned Aerial Vehicle: Applications in Agriculture and Environment

Springer

Editors
Ram Avtar
Faculty of Environmental Earth Science
Hokkaido University
Sapporo, Hokkaido, Japan

Teiji Watanabe
Faculty of Environmental Earth Science
Hokkaido University
Sapporo, Hokkaido, Japan

ISBN 978-3-030-27156-5 ISBN 978-3-030-27157-2 (eBook)
https://doi.org/10.1007/978-3-030-27157-2

This Springer imprint is published by the registered company Springer Nature Switzerland AG.
The registered company address is: Gewerbestrasse 11, 6330 Cham, Switzerland

Preface

In recent years, unmanned aerial vehicles (UAVs) are being extensively used in many applications, namely, not only environmental science but also precision agriculture, intelligent transportation, Internet of Things (IoT), surveillance, security, rescue operations, and entertainment industry. A UAV equipped with a multiple of sensors, each of which providing a piece of complementary information, augments the data collected using ground-based sensors and remotely sensed imagery captured from satellites. This book, *Unmanned Aerial Vehicle: Applications in Agriculture and Environment*, provides conceptual as well as operational aspects of unmanned aerial vehicles and their role in agriculture and environment. This book is not meant for engineers who want to design unmanned aerial vehicles for their applications; rather, the key focus of this book is on the use of UAVs in applications including precision agriculture, forest management, crop monitoring, fertilization management, plant health monitoring, area management, road patrolling, 3D modeling, topographic surveys, etc. The readers interested to dwell into design aspects of UAVs are encouraged to refer to books specifically written on design aspects of UAVs. The book comprises a total of 14 chapters in which each chapter is self-contained, providing some aspect of UAV in agriculture, environment, or an allied field. Several case studies have been included in the book enabling the readers to discover and explore as the research develops in this field. This book is intended for use by undergraduate and graduate students as well as by professionals who use UAVs in their respective applications. This book is written in such a way so that the readers can skip or skim through the chapters in which they are not interested without losing the context. Any suggestions or comments on the material of this book are welcome. We would like to sincerely thank the publishers of this book and all the contributing authors who adhered to the time frames in the publication process. We would like to thank the Graduate School of Environmental Science/ Faculty of Environmental Earth Science for providing the research facilities as well as the Office for Developing Future Research Leaders (L-station) of Hokkaido University for the financial support. We are also thankful to the Global Land

Programme (GLP) Japan Nodal Office for the encouragement and support. We would also like to acknowledge the time and effort spend by our research students in proofreading the contents of this book. We wish the readers all the best.

Sapporo, Japan

Ram Avtar
Teiji Watanabe

Contents

Chapter 1
Introduction

Ram Avtar and Teiji Watanabe

Abstract This chapter gives an introduction to the brief history, development, and state-of-the-art technology used in unmanned aerial vehicles (UAVs) with their application in the fields of agriculture and environment. The chapter also outlines the structure of book consisting of several chapters, each of which is authored by researchers working in their respective fields throughout their research career. The contributing authors are an amalgam of academicians, researchers, industry professionals, and users of such technology. The research areas considered in the case studies of this book cover diverse land from different countries and regions. This chapter summarizes the structure and approach toward the use of UAVs in an attempt to tackle the various challenges and issues faced in the use of UAVs.

Keywords UAV · Remote sensing · Precision agriculture · Drone · Multi-sensor · 3D model

1.1 Introduction

Unmanned aerial vehicle (UAV) systems were primarily used for military purposes and were later used for civilian applications. Historical UAV systems had limited control over their trajectory because of lack of sophisticated sensor technology. *Kettering Bug* and *Sperry-Curtis Aerial Torpedo* are the two historical UAVs used in military and navy applications in 1917. In 1937, a series of UAVs known as RP-1, RP-2, RP-3, and RP-4 were developed by Walter Righter and Kenneth Case (Fahlstrom and Gleason 2012). In the modern era, due to availability of differential GPS and sophisticated cameras, UAVs are widely used in many aspects of life.

New and emerging technology like UAV is trying to provide solutions to some unresolved problems of societal and the environmental. Unmanned vehicles can be classified into five different types according to their operation, viz., unmanned

R. Avtar (✉) · T. Watanabe
Faculty of Environmental Earth Science, Hokkaido University, Sapporo, Hokkaido, Japan
e-mail: ram@ees.hokudai.ac.jp

R. Avtar, T. Watanabe (eds.), *Unmanned Aerial Vehicle: Applications in Agriculture and Environment*, https://doi.org/10.1007/978-3-030-27157-2_1

ground vehicles, unmanned aerial vehicles, unmanned surface vehicles (operating on the surface of the water), unmanned underwater vehicles, and unmanned spacecraft. Unmanned vehicles can be either remotely guided or autonomous vehicles. The technology has found wide usage in agriculture, industry, transportation, communication, surveillance, and environment applications. It is widely used in precision agriculture for crop health monitoring, crop yield, and damage assessment. Apart from agriculture, it is also used in archaeological prospects and maintenance of traffic control in public gatherings to reduce the risk of stampede. In the recreational sector, UAV technology is being widely used in photography and moviemaking for special effects.

Like robotic technology, the use of UAVs has pros and cons, and their use for an application depends on the region, weather conditions, and national regulations. An unrestricted use of UAV might lead to more damage to the society than its benefits. With the control of local government and cooperation of citizens, UAVs equipped with RGB cameras, thermal cameras, multispectral sensor, hyperspectral sensor, and Light Detection and Ranging (LiDAR) sensor along with Global Positioning System (GPS) can be used in agriculture, health, transportation, and urban planning applications (Toro and Tsourdos 2018).

Various cameras equipped on to the UAVs give us information about crop yield, soil quality, and weed infestation in any part of the agriculture field. It is also used in flood and drought detection to help the government in relief operations. The role of UAVs in measuring forest carbon and observing phenomenon like glacier melting has proven to be a major step toward environment preservation. UAVs have been shown promising contributions in remote sensing data collection for agricultural and forestry applications due to their low cost, lightweight, and low airspeed aircraft capability. Unlike high-altitude aircrafts and satellites, UAVs can operate unnoticed and below the clouds. Aerial photographs captured by the UAV can bridge the gap between ground-based observations and remotely sensed imagery captured with conventional aircraft or satellite platforms.

This book mainly targets applications of UAVs to agriculture and environment in different parts of the world. Each chapter gives a brief introduction to one or the other area of UAV-based remote sensing along with details of research methodologies used in the respective areas. The demonstration of the methods and tools deployed in collecting primary and secondary data makes novice readers as well as professionals working on UAVs find this book easy to understand and useful. The book also discusses present challenges for the use of UAV in civil applications. The book also discusses latest research trends of UAV usage and its prospects. The UAV applications covered in this book include precision agriculture, forest management, fertilizer management, crop yield estimation, plantation management, disease detection, volcanic crater monitoring, road safety, and security and surveillance. The contributing authors of this book are from various research fields who have contributed significantly either to UAV technology or to the application of UAVs in research. The book also discusses opportunities and strengths related to UAV research in Asia.

1.2 Structure of the Book: Unmanned Aerial Vehicle—Applications in Agriculture and Environment

We do believe that this book will prove to be useful to those readers who wish to start their research career in UAV technology and also to professionals who want to explore more in UAV technology. The rest of the chapters of this book are organized as follows.

Chapter 2 provides insights into use of UAVs in precision agriculture. The traditional farming methods adopted in developing countries not only bring low productivity and crop yield, but also cause degradation of resources. The use of geographic information and communication technology (Geo-ICT) in precision agriculture helps for better yield (Tokekar et al. 2016). Drone-based multi-sensor system consisting of various sensors and cameras captures various parameters of soil and crop conditions leading to an effective way to farm management. Crop health monitoring based on leaf area index (LAI), normalized difference vegetation index (NDVI), photochemical reflectance index, crop water stress index, and similar other indices is carried out in a better manner with the use of UAVs. The role of UAVs in identifying biochemical stress in real time helps to carry out effective corrective measures.

Chapter 3 discusses the roles of UAVs in precision agriculture and forest management using multispectral imaging system. Keeping in view the unsuitability of using conventional push-broom spectrometers in UAVs, sequential two-dimensional spectral imager is used in UAVs for their inherent advantages. Several image processing and analysis tools used for hyperspectral images are also elaborated in this chapter. A pipelined processing technique has been discussed which makes the processing of hyperspectral images much faster than conventional methods meant for intensity level images. Several applications of UAV-based sensors for acquiring hyperspectral images in precision agriculture have also been discussed.

Chapter 4 deals with the use of UAV technology for fertilizer management in grassland of Hokkaido region, Japan. Overfertilizing agriculture areas has negative effects on water quality and affects the health of animals and human beings who use it. The aim of this study is to use multi-temporal UAV data to identify overfertilized areas. It is difficult to get high-resolution satellite data in the study area because of cloud cover. Hence, UAV techniques are used to fill the data gap which proved very helpful. Processing of data includes calculation of blue normalized difference vegetation index (BNDVI), which helps to detect possible overfertilized areas. Analysis for BNDVI intensity and comparison of UAV ortho-image and planet satellite images are also discussed.

Chapter 5 discusses corn crop yield prediction through corn plant height estimation generated by 3D photogrammetry based on structure from motion (SfM) technology (Javernick et al. 2014). Multi-temporal UAV data were collected along with ground data to develop empirical an model between ground calculated yield and its relationship with corn plant height. The UAV-derived normalized difference vegetation index (NDVI) data shows high sensitivity in the early growth stages of corn. As the estimated average height measured using UAV saturates 4 weeks later than that of UAV measured NDVI, it is more appropriate to monitor plant growth with respect to height instead of NDVI.

Chapter 6 discusses use of UAV and geographical information system (GIS) to support oil palm replanting programs in Malaysia. In 2018, the global palm oil consumption reached 30% of the global vegetable oil, turning it into the most consumed vegetable oil. Industries have been trying to satisfy the increasing global vegetable oil demand mainly by expanding area, which led to environmental concern about deforestation and biodiversity loss in tropical forests. To improve oil palm yield per hectare, implementation of best agricultural practices has been set as one of the RSPO's (the Roundtable on Sustainable Palm Oil) eight principles, in which both prior and post planting operations are being considered (RSPO 2013). However, as oil palm is cultivated in large-scale plantations, these operations are laborious and time-consuming. In the following chapter, the authors demonstrated a general work flow for oil palm tree detection and height measurements using UAV data. The use of local maximum (LM) and template matching (TM) algorithms are explained in detail.

Chapter 7 describes the importance of UAV technology in oil palm plantation area and health management. The UAV data of oil palm plantation areas in Malaysia were collected to acquire information about the plantation management practices, as well as oil palm health assessment. The UAV data are useful in plantation land management and identification of suitable sites for plantation. The precise information on landscape can reduce the cost for development of oil palm plantation area. The results show that multispectral camera is useful to identify the plant stress and detection of stunted tree crown.

Chapter 8 describes the role of unmanned aerial vehicle system (UAVS) technology in bridging the gap between field data collection and satellite-based observations. In this chapter, the authors shared their experiences of UAV applications in forest and plantation operation in Sabah and Sarawak, Malaysia. The benefits of UAV applications are improved efficiency by fast and timely geospatial data acquisition for various operational as well as for management purposes. This is possible by current advancement of consumer drones and availability of open-source UAVS technology for custom-made UAVS. The authors have discussed various case studies in forestry and plantation operations such as infrastructure management, roads, nursery, boundary, and encroachment monitoring.

Chapter 9 deals with the application of UAV-based SfM-MVS (structure from motion-multi-view stereo) mapping the cliff face in the central crater of Miharayama, Izu Oshima, central Japan (Chiba et al. 2019). The UAV can be used in many applications where conditions are not favorable or dangerous. A measurement of active volcano is one such application. UAV can avoid human damage as well as acquire ortho-image with higher resolution promptly as compared to satellite data in volcanic study. It is essential to predict the lava overflow time by observing the height of lava lake surface only. In this chapter, the authors used an UAV to scan in and around the Mt. Miharayama central crater in Izu Oshima Island and developed the 3D model by applying SfM-MVS technology. The relationship between altitude of lava lake surface and crater volume using the 3D model was calculated.

Chapter 10 discusses the history, development, and prospects of UAV-based remote sensing to topographic and vegetation surveys. A study on estimation of volumetric changes in a sea cliff of a peninsular rock in Japan known as Suzume Jima Island using UAV and terrestrial laser scanning (TLS) has been presented in this chapter. Structure from motion-based multi-view stereo technology is used to construct a 3D model of forestry from 2D images captured with the help of UAVs. Apart from constructing a 3D model, digital surface model (DSM), digital terrain model (DTM), and digital canopy model (DCM) are also created from the aerial images. These models enable researchers to estimate tree height for various applications in remote sensing. The use of the method for survey of coral reefs in marine environment and for measurement of snow depth distribution has also been discussed.

Chapter 11 deals with augmentation of road safety operations using UAVs equipped with thermal cameras. With many vehicles on the road in urban cities, road mishaps are rising at an alarming rate. This is partly because of poor visibility conditions on roads due to fog, smog, uncontrolled environment conditions, and presence of obstacles on the road. Although modern automobiles are equipped with several safety devices, additional devices like thermal camera mounted on UAVs can detect obstacles on the road and assist drivers in prevention of accidents for their own safety as well as for the safety of pedestrians. This chapter presents details of hardware modules and software routines required for the task so that the rescue operations team can be called for rescue to avoid causalities.

Chapter 12 addresses various issues for enhancement and fusion of multispectral images. At the outset, the need for enhancement and fusion of multispectral images is outlined, followed by methods available for their enhancement and fusion. The advantages and disadvantages of each method are also presented. Various types of distortion occurring in multispectral images due to sensor noise, environmental changes, and interference effects have been discussed. Several commercial and open-source software available in the market for processing multispectral imagery data have been discussed along with their salient features. The major challenge of using multispectral images for various applications is their huge size due to the presence of several bands captured at different wavelengths. The computational complexity of methods used for processing of multispectral images makes their use limited for online applications. The chapter focuses on processing methods which are computationally fast but at the same time have significant performance.

Chapter 13 discusses the application of UAV in urban green space (UGS) mapping. In developing countries, cities witness special dynamics of urban transition with uneven demographic densities, changing landscape patterns, traffic and congestion, and other environmental challenges. Geospatial data of UGS are mostly unavailable for emerging Indian cities. As use of UAVs in urban areas is still limited, the study recommends more experimentations and trials to arrive at set methodology which could help in mapping of UGS and other qualitative data gathering to support urban planning and urban greening. A case study of Nagpur city is presented to highlight specific direct applications. The review finds UAVs as a cost-effective and efficient tool with high-resolution images that are useful for planners and decision-makers. While regulation hides its wide applicability, the cost component, flexibility, timely monitoring, and accessibility weight UAVs as a suitable tool for data collection in urban areas.

Chapter 14 deals with the risk and challenges associated with the emerging technology to be deployed in Indian agriculture. It is highly diversified with varied physical features and sociocultural practices of the agriculture community. The paper draws empirical results from in-depth interviews carried out as a part of the primary survey based on the snowball sampling technique. Accordingly, it advances in adhering to the responsible deployment of the technology and ushering accountability in governance to enhance civil UAV innovations in the crop insurance application.

To summarize, use of UAVs has increased rapidly in various sectors like transportation, agriculture, environment, disaster studies, etc. This is due to the fact that data captured by UAVs is quite valuable for many applications. The only bottleneck in the use of UAVs is immense data which needs faster processing units. Deep learning has given us the tool to process a huge amount of data with many interesting applications. We wish all the readers good luck.

References

Chiba T, Urayama T, Mochizuki T, Miura S, Naruke S (2019) Making a 3D model of a crater using UAV. For the future 2019 Asia air survey, pp 92–93

Fahlstrom P, Gleason T (2012) Introduction to UAV systems. Wiley, Hoboken

Javernick L, Brasington J, Caruso B (2014) Modeling the topography of shallow braided rivers using structure-from-motion photogrammetry. Geomorphology 213:166–182

RSPO (2013) RSPO principles and criteria for sustainable palm oil production. https://rspo.org/key-documents/certification/rspo-principles-and-criteria

Tokekar P et al (2016) Sensor planning for a symbiotic UAV and UGV system for precision agriculture. IEEE Trans Robot 32(6):1498–1511

Toro FG, Tsourdos A (2018) UAV-based remote sensing colume-2, special issue. MDPI sensors, ISSN:1424-8220

Chapter 2
Precision Agriculture and Unmanned Aerial Vehicles (UAVs)

Rahul Raj, Soumyashree Kar, Rohit Nandan, and Adinarayana Jagarlapudi

Abstract Farming in developing countries is majorly dependent on the traditional knowledge of farmers, with unscientific agricultural practices commonly implemented, leading to low productivity and degradation of resources. Moreover, mechanization has not been integral to farming, and thus managing a farm is a time-consuming and labor-intensive process. Consequently, precision agriculture (PA) offers great opportunities for improvement. Using geographic information and communication technology (Geo-ICTs) principles, PA offers the opportunity for a farmer to apply the right amount of treatment at the right time and at the right location in the farm. However, in order to collect timely high-resolution data, drone-based sensing and image interpretation is required. These high-resolution images can give detailed information about the soil and crop condition, which can be used for farm management purposes. Leaf area index, normalized difference vegetation index, photochemical reflectance index, crop water stress index, and other such vegetation indices can provide important information on crop health. Temporal changes in these indices can give vital information about changes in health and canopy structure of the crop over time, which can be related to its biophysical and biochemical stress. These stresses may have occurred due to insufficient soil nutrient, inappropriate soil moisture, or pest attack. Through UAV-based PA, stressed areas can be identified in real time, and some corrective measures can also be carried out (e.g., fertilizer and pesticide spraying). Moreover, the advantages and different approaches to integrate the UAV data in the crop models are also described.

Keywords Drone-based sensors · Vegetation indices · Agricultural drones · UAV-based precision agriculture cycle

---

R. Raj
Indian Institute of Technology Bombay – Monash Research Academy, Mumbai
Maharashtra, India
e-mail: rahul_raj@iitb.ac.in

S. Kar · R. Nandan · A. Jagarlapudi (✉)
Indian Institute of Technology Bombay, Mumbai, Maharashtra, India
e-mail: karsoumya@iitb.ac.in; rohitnandan@iitb.ac.in; adi@csre.iitb.ac.in

R. Avtar, T. Watanabe (eds.), *Unmanned Aerial Vehicle: Applications in Agriculture and Environment*, https://doi.org/10.1007/978-3-030-27157-2_2

2.1 Precision Agriculture

The Industrial Revolution has pushed agriculture practices toward greater energy inputs through the use of big machinery, chemicals, and fertilizers. However, these practices may lead to low soil fertility, soil erosion, soil salinization, compaction of subsoils, and soil-water pollution, which are having negative societal and environmental implications (Liaghat and Balasundram 2010). Precision agriculture (PA) is an innovative and integrated farming approach which enables farmers to use evidence-based decision-making at the farm level, to ensure optimal use of resources to minimize such societal and environmental implications (Tokekar et al. 2016). PA can use traditional knowledge together with spatial information and management-intensive technologies. This helps in making the system sustainable, productive, and profitable. The technologies frequently used in PA include Geographic Information System, Global Positioning System, remote sensing, computer modelling, variable rate technology, and machine learning approach with advanced information processing for timely crop management (Liaghat and Balasundram 2010). The PA cycle can be explained in the following steps:

1. Data collection: Within-field variability in soil and crop parameters and local weather conditions are measured, monitored, and mapped.
2. Data interpretation: Data interpretation using various crop models or image interpretation and/or data assimilation techniques are undertaken to identify spatially variable parameters.
3. Application: Based on data processing results, farm management can be tailored for the right place, the right time, and the right amount.

Precision agriculture cycle is shown as in Fig. 2.1.

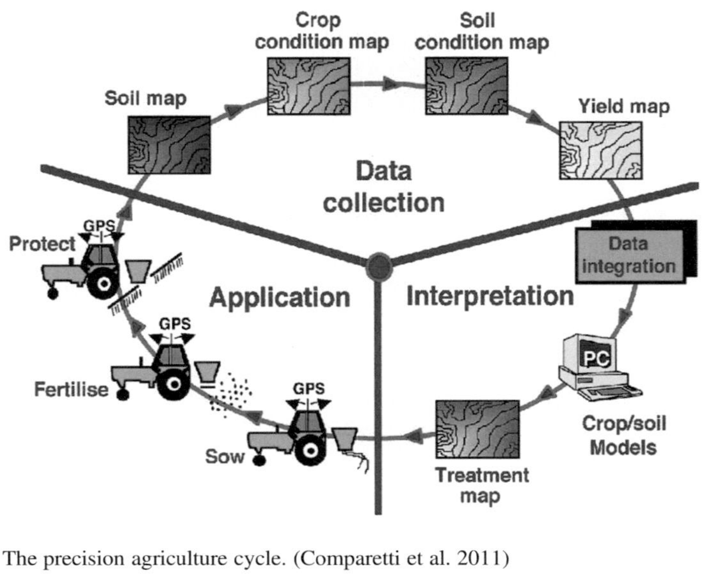

Fig. 2.1 The precision agriculture cycle. (Comparetti et al. 2011)

Data collection and database generation is a very important part of PA, requiring wise selection of sensors and their deployment in the farm to ensure accuracy and precision of on-farm decision using a decision support system. Sensors can be on ground or airborne, with the purpose of finding information such as the health of the crop, its growth stage, physical and chemical properties of the soil/pant, temporal meteorological data, etc. Remote sensing and geospatial techniques are an integral part of the data collection and processing process and are used to detect in-field variation. High-resolution satellite imagery is very useful for studying variations in crop and soil conditions. However, problems associated with the availability and cost of such imagery at an appropriate spatial and temporal resolution suggest that an alternative, such as "small unmanned aerial systems (UAS)," is needed for operational PA (Zhang and Kovacs 2012).

2.2 Satellite vs Drone Sensing

Satellite-based remote sensing is one of the traditional methods to acquire remotely sensed data, but the freely available images can only help in getting data at a resolution of 30 m or greater, which is too coarse for many applications. Some commercial satellites provide sub-meter resolution satellite imagery (spatial resolution < 1 m for panchromatic and > 1 m for multispectral) for a given time and place/area at a given price but are typically infrequent in time (Liaghat and Balasundram 2010). While satellite images may be the best option in the case of very large areas, the coarse spatial and temporal resolution with long revisit durations significantly limits its application, particularly during cloudy conditions of the atmosphere. Consequently, images taken by low-altitude remote sensing platforms such as small unmanned aerial vehicles or small manned aircraft provide an alternative.

The cost of operation is low for UAVs, and they can map areas with very high spatial resolution and at desired temporal repeat (Zhang and Kovacs 2012). Moreover, they can fly below the cloud, maximizing data availability. However, there are other problems which affect drone operation, including a high windy condition in which the drone can't be flown, higher risk of a crash in case of operator error, sudden weather change or loss of power, and low battery life which limits spatial coverage. Moreover, the commercial satellite takes usually 7 days to provide processed images, while drone-based images can be processed in near real time (Satimagingcorp 2018). While drones can be flown at any time as per requirement with minimal operational cost, for a small- or medium-scale farmer, it is very difficult to get commercial satellite images as the cost is high and the minimum area which can be ordered is the range of 25 km^2 (Landinfo 2018). Another advantage of drone data is that canopy images can be taken at various view angles in order to analyze the structure of the canopy. The limitations associated with satellite data over drone-based high-resolution images are summarized in Table 2.1.

The type of drone which can be used for agricultural application may depend on the size of the farm and type of cameras/sensors which need to be used. Battery

Table 2.1 Satellite data vs drone-based sensing

Sub-meter resolution commercial satellite image	Drone-based high-resolution image
Cloud cover and atmospheric dust particles create a bottleneck on image acquisition	As flight height is low thus limited effect of cloud cover
Real-time image acquisition and processing are not possible and usually take 7 days of delay	Images can be obtained and processed in a few hours depending on the size of the farm
Images captured at some fixed time of day depending on the frequency of revolution of the satellite	Images can be captured at the desired time of day
Maximum available panchromatic geometrical resolution is 30 cm, while multispectral resolution would be 1.24 m	Spatial resolution may go around 1 cm
Minimum area map which can be ordered is 25 square km or more (if only natural color map is required, then 10 square km)	Map can be generated for a small and medium area which would be much cheaper than satellite imaging
Images generally are taken from zenith	Images at a different angle can be taken which will help in getting architectural information of canopies

capacity also plays an important parameter which decides flight time of a drone. Table 2.2 shows the types of drones and their capability to carry various sensors.

2.3 UAV-Based Sensors in Precision Agriculture

Applied agricultural research is generally related to productivity improvement, yield quality enhancement, cost-effective technologies, selection of better crop genotypes, and weather-resistant crops. There have been a significant number of studies conducted in this domain. However, more than 50% (~3000) of the papers available on drone use in agriculture are published after 2016, with almost 90% of the papers published after 2013 (Web of Science). This is due to improved efficiency of UAVs and scanning imagers over the last 5 years, together with dissatisfaction with satellite images for making on-farm decisions in near real time (Matese et al. 2015). Since satellite's spatial resolution is coarse for small farm and ground-based sensors cannot cover a big area on the field, thus drone-based data collection shows an alternative path to collect data at such temporal and spatial resolution.

UAVs have the capability to carry various sensors which are useful for studying crop-related parameters. Literature shows that drones can be integrated with optical sensors like RGB camera, multispectral camera, and hyperspectral camera. Thermal cameras can also be used on drones which help in identifying water stress in the crop (Calderón et al. 2013). LiDAR sensors are used on drones to estimate the height of the canopy which ultimately helps in biomass estimation of the crop. Apart from digital data collection, drones are also applied to do aerobiological sampling above agricultural fields for early identification of pest attack on the crop (Schmale et al.

Table 2.2 Types of drones and their capability to carry sensor

Type	Weight in kg (including payload)	Types of sensors which can be mounted on the drone	Area coverage capacity of the drone
Nano	Less than 0.25 kg	This has not been used in agriculture till now as sensors are usually of more weight and cannot be lifted by nano-drones	NA
Micro	0.25 < weight < 2	Small RGB, lighter multispectral camera, and small LiDAR sensor can be mounted on this drone	Can cover up to 4–5 acres of ground area depending on the height of flight. Flight height is generally kept lower than 100 m
Small	2 < weight < 25	High-resolution RGB camera, multispectral camera, LiDAR sensor, lightweight hyperspectral imager, and small thermal imager can be mounted on the drone	Can cover up to 10–20 acres of ground area depending on the height of flight
Medium	25 < weight < 150	Bigger high-resolution RGB camera, multispectral camera, LiDAR sensor, medium-weight hyperspectral imager, and thermal camera can be mounted on the drone. It can also be used for spraying of pesticides	Can cover up to 100 acres of ground area depending on the height of flight. Flight height is generally higher than 50 m
Large	Greater than 150	Bigger and heavyweight cameras and sensors can be mounted on the drone. It can be used for spraying of pesticides	Can cover more than 100 acres of ground area. Flight height is generally higher than 100 m

2008). Drones can also be used to take corrective measures in farms like the spraying of pesticides in stressed areas of the farm. Table 2.3 shows the different sensors which can be installed on the drone to study various characteristics of crops.

2.3.1 *Visible and IR Imagers and Sensors (400–2500 nm)*

Imagers in visible and IR spectrum are popularly used from a drone platform for vegetation mapping. These optical sensors can be used for estimation of biomass, LAI, identification of various growth stages, and healthiness of a crop. Pest identification, survey of a farm, mapping, etc. are also done using these sensors. Below is the list of optical sensors available for drone-based sensing.

(a) **RGB Camera**

A digital RGB camera can be mounted on a UAV and top of the canopy or stereo images of farm can be captured. These images are in visible bands (400–700 nm)

Table 2.3 Drone-based sensors and their use

Drone-based sensing of vegetation The following cameras/sensors/instruments can be installed on a drone, and data can be collected which can be further used for estimation of various crop parameters:
1. PAR and IR imagers and sensors (400–2500 nm) (a) RGB camera (b) Multispectral camera/red-edge camera (c) Hyperspectral imager (400–2500 nm) (i) Snapshot imager (ii) Line scanner imager
2. Thermal camera (3000–12,000 nm)
3. LiDAR sensor
4. Aerobiological sampling
5. Spraying of pesticides through drone

which collect reflectance in three broad ranges of wavebands: red, green, and blue bands. The spatial resolution of the image depends on camera specifications and height from which drone was flown. With good flight planning, these cameras are capable of collecting very high spatial resolution images which might give pixel resolution up to 1 mm. However, the spatial resolution is decided as per objective of the work, and collecting extremely high spatial resolution data might not be a good idea as it will require more space to store and greater time to process.

These images can be used for making ortho-mosaic of the farm, studying the structural properties of the plants/trees (RAMI), detecting the weed location or pest-affected areas in the farm, and estimating the LAI of the crop. Height estimation of crop is also possible from drone-based RGB images which helps in biomass estimation. These cameras are cheaper than multispectral or hyperspectral imagers, easy to operate, lighter in weight, and thus very popular for vegetation studies.

(b) **Multispectral Camera/Red-Edge Camera**

Multispectral cameras consist of 4–6 bands of around 10–50 nm bandwidth in blue, green, red, red-edge, and NIR regions of the electromagnetic spectrum. These cameras are generally used to calculate normalized difference vegetation indices and are capable of estimation of biomass and identification of highly stressed areas in the farm.

(c) **Hyperspectral Imager (400–2500 nm)**

(i) **Snapshot imager:** These imagers are capable of acquiring images in several narrow bands from visible to IR region of EM spectrum. Bandwidth is generally around 10 nm (broader than line scanner imagers). These imagers are easy to handle (compared to line scanner), and data is relatively easy to process as raster file is generated by the imager which can be directly used in GIS software.

(ii) **Line scanner imager:** These imagers are comparatively complex to operate. Speed of the drone is decided by the camera frame rate so drone should fly as per frame rate of the camera as the camera captures a narrow row on the ground in a given time, and thus flight should be synchronized accordingly. Bandwidth can be narrow (around 2 nm). These imagers are relatively difficult to handle and collected data is complex to process.

2.3.2 *Thermal Camera (3000–12,000 nm)*

Thermal cameras are very useful in determining water stress in the crop. It has been seen that the crops that have water stress are relatively at higher temperature than the crops which are not having water stress. However, this temperature difference is time dependent (morning, afternoon, or evening) and very much affected by varying solar radiations. The temperature difference captured by the thermal camera can distinguish the water-stressed crop easily during afternoon when sky is clear and solar radiation is available (Bellvert et al. 2014).

2.3.3 *LiDAR Sensor*

Light Detection and Ranging (LiDAR) sensor is very useful in measuring canopy height. The sensor can collect data from up to 250-m height (depending on manufacturer), and the accuracy may be of few millimeters. This sensor data is generally fed to a photogrammetry software for analysis purpose.

2.3.4 *Aerobiological Sampling*

An air sampler can be fitted on a drone which can collect and store air samples above agricultural farms. The analysis of these aerobiological samples helps in early identification of pest attack on crop (Schmale et al. 2008).

2.3.5 *Spraying of Pesticides Through Drone*

This is a corrective measure which can be implemented through the drone. Tankers filled with pesticide are carried by drones and the spraying can be done precisely in those areas which are found to be stressed.

2.4 Vegetation Indices and Other Techniques for Drone-Based Data Analysis

From remote sensing techniques, different vegetation indices and soil properties can be calculated using various airborne sensors. For example, leaf area index and NDVI are two popular indices used to indicate crop health/state, while thermal cameras can be used to estimate water stress (Berni et al. 2009). Canopy reflectance can be used to identify various biophysical and biochemical properties of the canopy through either a physical or data-driven model such as machine learning model or machine learning technique that is evolving very quickly and has shown better results in many cases. To predict the crop health status efficiently, it is very important to collect reliable farm data which represents the farm at sufficient spatial and temporal resolution.

Use of vegetation indices for estimating various crop biophysical and biochemical characteristics is one of the popular methods. However, there is always a sensitivity issue associated with indices, e.g., LAI or NDVI tends to saturate with the increasing amount of biomass in the area. Indices have shown good results with satellite data and with reasonable classification accuracy. Let us discuss some popular indices in this chapter.

Leaf area index (LAI) is a parameter associated with the physiological processes of the crop. It is used to study growth, photosynthesis, and transpiration of plants and to know interception of radiation in the canopy. LAI is also used in crop yield prediction and water balance modelling, defined as the total one-sided area of photosynthetic tissue per unit ground surface area (Jonckheere et al. 2004). When LAI value increased from approximately 3–4 (depending on the canopy), then NDVI loses its sensitivity toward change in LAI and starts saturating. This is because chlorophyll is a highly efficient absorber of red radiation, and thus after some point adding more chlorophyll to the canopy or in other words increasing leafy material in the canopy will not change red reflectance much. To overcome this situation, several solutions have been developed. One of the simplest solutions is to use wide dynamic range vegetation index, i.e., WDRVI. To make this index, weighting factor ranging from 0 to 1 is used with NIR reflectance (in the numerator as well as in denominator) in the formula of NDVI (Gitelson 2004):

$$\mathrm{NDVI} = \frac{\mathrm{NIR} - R}{\mathrm{NIR} + R} \,\&\, \mathrm{WDRVI} = \frac{a * \mathrm{NIR} - R}{a * \mathrm{NIR} + R}. \tag{2.1}$$

When weighting factor approaches to 0, the linear relationship between WDRVI and LAI graph tends to increase but with the reduction in sensitivity of LAI changes in sparse canopies. Another index which shows better sensitivity with LAI is designed with blue bands. This is called enhanced vegetation index:

$$\mathrm{EVI} = 2.5 * \left(\frac{\mathrm{NIR} - R}{\mathrm{NIR} + (6 * R) - (7.5 * B) + 1}\right). \tag{2.2}$$

Later EVI is modified and blue band is removed. The modified version of EVI is known as EVI2. Apart from having a linear relationship with LAI, EVI2 has less soil sensitivity compared to NDVI:

$$\mathrm{EVI2} = 2.5 * \frac{\mathrm{NIR} - R}{(\mathrm{NIR} + (2.4 * R) + 1)}. \tag{2.3}$$

LAI can be combined with other vegetation indices to achieve maximal sensitivity. For maize-soybean rotation crop, calculated gLAI for maize and soybean ranged from 0–6.5 to 0–5.5, respectively. For gLAI lower than 2, NDVI is most sensitive, while for gLAI greater than 2, simple ratio (SR = NIR/*R*) and chlorophyll indices (CI = NIR/*G* − 1) are most sensitive. However, this relationship is crop specific and may change with other crops. The best index combination for maize and soybean is combination of NDVI and SR. With this combination, coefficient of variance for maize and soybean was less than 20% and 23%, respectively (Nguy et al. 2012). In an experiment on grapevines, it is seen that NDVI values calculated through UAV-based camera show strong linear correlation ($R^2 = 0.97$) with NDVI calculated by field spectroradiometer (Primicerio et al. 2012).

There are various index-based and sensor-based methods to estimate crop water stress. Thermal cameras can also be used to estimate crop water stress, and CWSI can be found as shown in Eq. 2.4 (Bellvert et al. 2014):

$$\mathrm{CWSI} = \frac{(Tc - Ta) - (Tc - Ta)\mathrm{LL}}{(Tc - Ta)UL - (Tc - Ta)\mathrm{LL}} \tag{2.4}$$

In Eq. 2.4, ($Tc - Ta$) is the canopy-air temperature difference, LL is the ($Tc - Ta$) values for lower limit, and UL is the upper limit of the same.

There is an index to estimate biomass of crop which is called "normalized green-red difference index (NGRDI)." NGRDI is found to be linearly related to biomass of alfalfa, corn, and soybean up to about 120 g m^{-2} (Hunt et al. 2005) as shown in Eq. 2.5:

$$\mathrm{NGRDI} = \frac{(\mathrm{Green\ DN} - \mathrm{Red\ DN})}{(\mathrm{Green\ DN} + \mathrm{Red\ DN})}. \tag{2.5}$$

UAV platforms are also being used to acquire fluorescence, temperature, and narrowband indices for water stress detection using a hyperspectral imager and a thermal camera. A strong relation is found among crown temperature, the blue-green BGI1 (R_{400}/R_{550}) index, and the chlorophyll fluorescence estimates (Zarco et al. 2012). In research done on olive orchards, it is observed that canopy temperature and physiological hyperspectral indices such as PRI and chlorophyll fluorescence are related with physiological stress caused by verticillium wilt (Calderón et al. 2013). Table 2.4 tabulates all the important vegetation indices compiled through a detailed literature.

Table 2.4 Indices used for identification of leaf water content and nitrogen content

	Indices	Formula	Source
Structural indices	NDVI (normalized difference vegetation Index)	$\text{NDVI} = \frac{\text{NIR}-R}{\text{NIR}+R}$	Jackson et al. (1980)
	WDRVI (wide dynamic range vegetation index)	$\text{WDRVI} = \frac{a*\text{NIR}-R}{a*\text{NIR}+R}\ (0 < a < 1)$	Gitelson (2004)
	RDVI (renormalized difference vegetation index)	$\text{RDVI} = \frac{R800-R670}{(R800+R670)^{0.5}}$	Roujean and Breon (1995)
	OSAVI (optimized soil adjusted vegetation index)	$\text{OSAVI} = \frac{(1+0.16)(R800-R670)}{(R800+R670+0.16)}$	Rondeaux et al. (1996)
	EVI (enhanced vegetation index used for LAI estimation)	$\text{EVI} = 2.5 * \left(\frac{\text{NIR}-R}{\text{NIR}+(6*R)-(7.5*B)+1}\right)$	Jiang et al. (2008)
	EVI2 (modified EVI) (less soil sensitive than NDVI)	$\text{EVI2} = 2.5 * \frac{\text{NIR}-R}{(\text{NIR} + (2.4*R)+1)}$	
Biomass estimation	NGRDI (normalized green-red difference index)	$\text{NGRDI} = \frac{(\text{Green DN}-\text{Red DN})}{(\text{Green DN}+\text{Red DN})}$	Hunt et al. (2005)
Chlorophyll indices	Red-edge reflectance index	$\frac{R750}{R710}$	Zarco-Tejada et al. (2001)
	DCNI (double-peak canopy nitrogen Index)	$\text{DCNI} = \frac{\frac{R720-R700}{R700-R670}}{(R720-R760+0.16)}$	Chen et al. (2010)
	TCARI (transformed chlorophyll absorption in reflectance index)	$\text{TCARI} = 3\left[(R700-R670) -0.2(R700-R550)\frac{R700}{R670}\right]$	Kim et al. (2002)
	Combined TCARI/OSAVI	$\frac{\text{TCARI}}{\text{OSAVI}}$	Haboudane et al. (2002)
	Carotenoid index	$\frac{R515}{R570}$	Zarco-Tejada et al. (2013)
Xanthophyll indices	PRI (photochemical reflectance index)	$\text{PRI} = \frac{R570-R539}{R570+R539}$	Gago et al. (2015)
	Normalized PRI	$\text{PRI norm} = \frac{R515-R531}{R515+R531}$	Gago et al. (2015)
Blue/green ratio index	BGI1	$\text{BGI1} = \frac{R400}{R550}$	Zarco-Tejada et al. (2012)

(continued)

Table 2.4 (continued)

	Indices	Formula	Source
	BGI2	$\text{BGI2} = \frac{R450}{R550}$	Zarco-Tejada et al. (2012)
Leaf equivalent water thickness	MSI (moisture stress index)	$\text{MSI} = \frac{R1600}{R820}$	Hunt et al. (1989)
	NDWI (normalized difference water index)	$\text{NDWI} = \frac{R860-R1240}{R860+R1240}$	Stimson et al. (2005)
	NDII (normalized difference infrared index)	$\text{NDII} = \frac{R820-R1600}{R820+R1600}$	Hardisky et al. (1983)
	MDWI (maximum difference water index)	$\text{MDWI} = \left[\frac{R\max - R\min}{R\max + R\min}\right]$ from $1500 - 1700$	Eitel et al. (2006)
Crop water Stress	CWSI	$\text{CWSI} = \frac{(Tc-Ta)-(Tc-Ta)\text{LL}}{(Tc-Ta)\text{UL}-(Tc-Ta)\text{LL}}$	Bellvert et al. (2014)

Apart from the index methods, hyperspectral drone data can also be analyzed based on its pixel-wise spectral signature. Various crops can be distinguished based on their characteristics of spectral signature. This might need machine learning approach to analyze the data. There are also physical models like PROSPECT, PROSAIL, LIBERTY, etc. that are available which take spectra as input and estimate plant's chemical and biophysical properties.

2.5 UAV-Based Precision Agriculture Cycle

Drones can be helpful in on-farm decision-making even before sowing starts. When farm soil bed is being prepared for sowing, UAV-mounted LiDAR sensor can be flown to check the flatness of the farm. If it is found that the soil bed is not flat enough, then based on elevation difference, the farm can be uniformly flattened. A uniformly flat field is one of the important requirements in order to stop unwanted movement of water in the farm. Research is going on to estimate soil nutrient content from drone-based sensors.

After sowing, temporal monitoring of the farms can be done using drone-mounted RGB camera. Images taken from these cameras can be used to monitor crop growth (biomass, LAI, height, etc.). The images can detect weed location in the farm and can also identify the pest-affected areas. In some crops like maize where tasseling happens, these images are capable of counting number of tassels which helps in early estimation of yield. During crop vegetative stage, apart from RGB, multispectral and hyperspectral cameras can also be used to not only estimate biophysical properties of the crop but also biochemical properties like leaf nutrient, water content, etc. Once location of these nutrient or water stressed areas, weeds, and

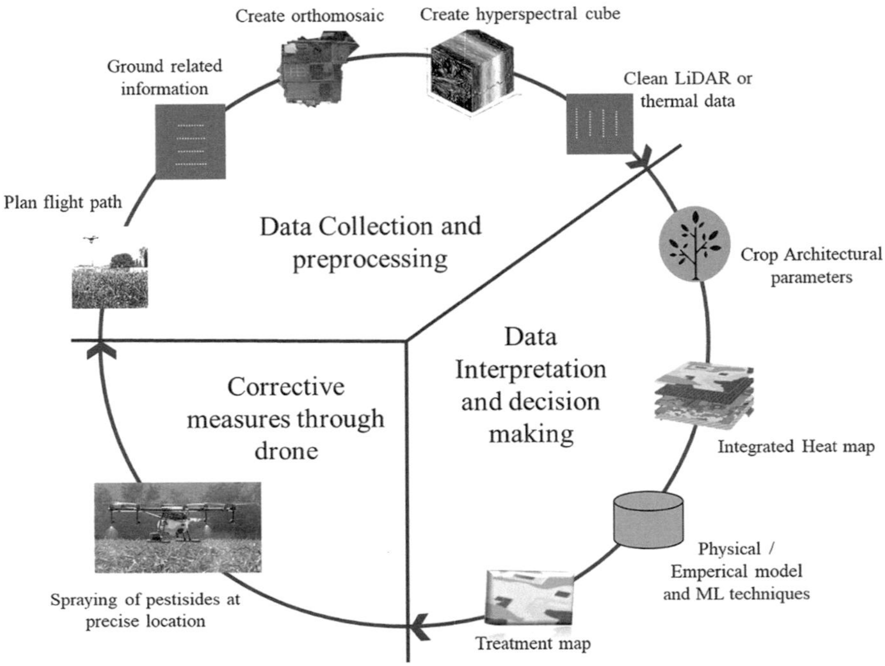

Fig. 2.2 The UAV-based precision agriculture cycle

pests are identified, drones can be used to take corrective measures and can spray pesticides or fertilizers or water precisely at the location in the farm. Figure 2.2 shows UAV-based precision agriculture cycle.

2.6 UAV in High-Throughput Plant Phenotyping

It has been observed that since the conceptualization of the process, phenotyping, there has been constant endeavor in evolving toward field phenotyping from lab phenotyping (Fukai and Fischer 2012). While several studies focus on modelling various phenotypic behaviors (Huang et al. 2010), several others experiment with different platforms that can both scale up and help understand the regional phenology better (Wallace et al. 2016). It's also stated in Wallace et al. (2016) that "the phenological development stage, however, can only be determined from an imagery collection rate that is unfeasible with aerial campaigns given the economic limitations of most natural resource budgets," which clearly distinguishes the edge of UAV imaging over other aerial remote sensing methods.

Although UAVs have been increasingly used for HTPP, the sensors on board differ with different applications. While RGB cameras are used for the morphological traits, multispectral/red-edge and hyperspectral cameras are used for the retrieval of biochemical traits. Holman et al. (2016) have used RGB camera-based UAV imagery to study both crop height and growth rate, by generating 3D digital surface models (via surface from motion, SfM, photogrammetry technique) from multi-temporal UAV data. The SfM model-derived estimates of the growth rate have been applied for the winter wheat field phenotyping experiment which contained 25 different varieties grown with 4 different nitrogen fertilizer treatments, and it could be identified that the per-day growth rates differed between 13 mm/day and 17 mm/day. Hence, UAV imaging not only helps in high throughput but also "precise" phenotyping. In a similar study by Watanabe et al. (2017), the NIR-GB camera-based UAV imaging has been exploited to derive height of the sorghum plants. These height estimates have subsequently been successfully used as the training data to the genomic prediction models for delineating the genotypic differences in sorghum, since the predicted and the ground truth plant height values were highly correlated ($r = 0.842$). There are also instances (Burud et al. 2017) where VIS/NIR multispectral cameras have been used to examine the efficiency of the UAVs as HTPP tools for plant breeding. In an attempt to delineate the field plots based on different vegetation indices obtained from UAV imagery, Haghighattalab et al. (2016) have developed a semiautomated image processing pipeline to perform the critical photogrammetric operations and radiometric calibration of the images. The relationships between vegetation indices (VIs) extracted from high spatial resolution multispectral UAV imagery and ground truth spectral data collected using handheld spectroradiometer have also been mentioned. Thus, UAVs enable faster and higher-resolution crop data collection (as part of HTPP) while simultaneously facilitating scientists and growers with improved precision agriculture practices on increasingly larger farms, e.g., site-specific application of water and nutrients.

Even though there are many advancements in UAV-based high-resolution data acquisition techniques, there is a need for research on image interpretation techniques to identify and estimate various crop physical and chemical properties like LAI, counting of crop kernels, leaf water and nitrogen percentage, etc. so that input resources can be utilized optimally.

2.7 Integration of UAV Data with Crop Models

There are two types of crop models. Empirical models are based on the regression relation between one or a few parameters and the observed data. Statistical models are less data intensive. The major limitation of these models is that they cannot be used for the regions and for environmental conditions for which historical datasets are not available (Jones et al. 2017). Process-based models simulate the crop physiological properties through time using differential equations to describe crop production. Within this conceptual umbrella, models employ functions, which approximate the hypothetical mechanistic canopy, and soil processes being simulated for a given time step (Wallach et al. 2014). The input data requirements of these models are more than empirical models (Di Paola et al. 2016). The process-based models can be used to explore the crop responses to climate change/variability condition and sets of management practices. This is the main advantage of the process-based models over empirical models.

Cropping systems operate at different spatial and temporal scales. The crop modelling requires the spatial distribution of soil characteristics (soil moisture), canopy state variables (LAI, biomass, nitrogen content, etc.), and meteorological data which are uncertain. The geospatial information of soil characteristics and canopy state variables can be estimated using remote sensing data.

In India, commonly, the farmers hold limited farm, and this invokes the heterogeneity in the farm. It is a challenging task to monitor the small farms accurately using satellite-based remote sensing data. For most of the agricultural applications, the physiological properties and high-resolution data are required, and these are the two major limitations in the satellite-based remote sensing. To overcome these two limitations, the crop models can be used to simulate the physiological properties, UAV can be used to collect the high-resolution remote sensing data, and both can be integrated. There are numerous researches carried out on the assimilation of satellite-based remote sensing and crop models (Jin et al. 2018). For accurate decision-making in heterogeneous farm, satellite sensing can be replaced by the drone sensing. To integrate remote sensing and crop models, the following approaches can be utilized:

In the first approach, the state variables of the models are directly replaced by the biophysical or biochemical products of the remote sensing data at each model's time step, viz., the remote sensing estimated LAI has been directly used to replace the state variables in the crop models (Schneider 2003; Hadria et al. 2006). This has reportedly improved the accuracy of the model's simulated crop production variables.

In the second approach, the biophysical characteristics or the biomass or crop yield estimated from drone-sensed data can be used to calibrate the crop model by adjusting its parameters. Several algorithms have used the calibration approach (Jin et al. 2018).

In the third approach, the data assimilation methods are used. Here, the simulated data from the crop model, continuously assimilated based on the assumption that better simulation data on the current simulation time step, will increase the accuracy of the simulation at next time steps. This method was found more accurate and popular in the past studies. There were numerous data assimilation algorithms such as 4DVar, EnKF, POD4DVar, ensemble square root filter, etc. used to integrate the state variables of crop models and remote sensing products such as soil moisture, AGB, and LAI (Jin et al. 2018).

Satellite remote sensing data have errors due to mix pixels, atmosphere, etc. These errors are inherent in the remote sensing products. In the first approach, crop models use remote sensing estimated products instead the simulated value, and the errors linked with remote sensing data directly affect the crop models' accuracy. This error could be reduced by using the drone sensing data instead of satellite remote sensing data. The second and third approach have more advantages and the optimization and assimilation algorithms are used to minimize the errors. If all three approaches are compared to each other, according to Jin et al. (2018), theoretically, the second method is better than the other two; however, the only drawback of the second method is that it requires many iterations for optimization, which results in increased computation time.

References

Bellvert J et al (2014) Mapping crop water stress index in a 'Pinot-noir'vineyard: comparing ground measurements with thermal remote sensing imagery from an unmanned aerial vehicle. Precis Agric 15(4):361–376
Berni JAJ et al (2009) Thermal and narrowband multispectral remote sensing for vegetation monitoring from an unmanned aerial vehicle. IEEE Trans Geosci Remote Sens 47(3):722–738
Burud I, Lange G, Lillemo M, Bleken E, Grimstad L, From PJ (2017) Exploring robots and UAVs as phenotyping tools in plant breeding. IFAC-PapersOnLine 50(1):11479–11484
Calderón R et al (2013) High-resolution airborne hyperspectral and thermal imagery for early detection of Verticillium wilt of olive using fluorescence, temperature and narrow-band spectral indices. Remote Sens Environ 139:231–245
Chen P et al (2010) New spectral indicator assessing the efficiency of crop nitrogen treatment in corn and wheat. Remote Sens Environ 114(9):1987–1997
Comparetti A et al (2011) Precision agriculture: past, present and future. Conference: international scientific conference "Agricultural engineering and environment. (Accessed from Researchgate)
Di Paola A, Valentini R, Santini M (2016) An overview of available crop growth and yield models for studies and assessments in agriculture. J Sci Food Agric 96:709–714
Eitel JUH et al (2006) Suitability of existing and novel spectral indices to remotely detect water stress in Populus spp. For Ecol Manag 229(1–3):170–182
Fukai S, Fischer KS (2012) Field phenotyping strategies and breeding for adaptation of rice to drought. Front Physiol 3:282
Gago J et al (2015) UAVs challenge to assess water stress for sustainable agriculture. Agric Water Manag 153:9–19
Gitelson AA (2004) Wide dynamic range vegetation index for remote quantification of biophysical characteristics of vegetation. J Plant Physiol 161(2):165–173
Haboudane D et al (2002) Integrated narrow-band vegetation indices for prediction of crop chlorophyll content for application to precision agriculture. Remote Sens Environ 81 (2–3):416–426
Hadria R, Duchemin BI, Lahrouni A, Khabba S, Er Raki S, Dedieu G, Chehbouni A, Olioso A (2006) Monitoring of irrigated wheat in a semi-arid climate using crop modelling and remote sensing data: impact of satellite revisit time frequency. Int J Remote Sens 27:1093–1117
Haghighattalab A, Pérez LG, Mondal S, Singh D, Schinstock D, Rutkoski J, Ortiz-Monasterio I, Singh RP, Goodin D, Poland J (2016) Application of unmanned aerial systems for high throughput phenotyping of large wheat breeding nurseries. Plant Methods 12(1):35
Hardisky MA, Klemas V, Smart M (1983) The influence of soil salinity, growth form, and leaf moisture on spectral radiance of spartina alterniflora canopies. Photogramm Eng Remote Sens 16(9):1581–1598
Holman F, Riche A, Michalski A, Castle M, Wooster M, Hawkesford M (2016) High throughput field phenotyping of wheat plant height and growth rate in field plot trials using UAV based remote sensing. Remote Sens 8(12):1031
Huang X, Sang T, Zhao Q, Feng Q, Zhao Y, Li C, Zhu C, Lu T, Zhang Z, Li M, Fan D (2010) Genome-wide association studies of 14 agronomic traits in rice landraces. Nat Genet 42(11):961
Hunt Jr, Raymond E, Rock BN (1989) Detection of changes in leaf water content using near-and middle-infrared reflectances. Remote Sens Environ 30(1):43–54

Hunt ER et al (2005) Evaluation of digital photography from model aircraft for remote sensing of crop biomass and nitrogen status. Precis Agric 6(4):359–378
Jackson RD et al (1980) Hand-held radiometry: a set of notes developed for use at the workshop of Hand-held radiometry. USDA, Oakland
Jiang Z et al (2008) Development of a two-band enhanced vegetation index without a blue band. Remote Sens Environ 112(10):3833–3845
Jin X, Kumar L, Li Z, Feng H, Xu X, Yang G, Wang J (2018) A review of data assimilation of remote sensing and crop models. Eur J Agron 92:141–152
Jonckheere I et al (2004) Methods for leaf area index determination. Part I: theories, techniques and instruments. Agric For Meteorol 121:19–35
Jones JW, Antle JM, Basso B, Boote KJ, Conant RT, Foster I, Godfray HCJ, Herrero M, Howitt RE, Janssen S et al (2017) Brief history of agricultural systems modeling. Agric Syst 155:240–254
Kim MS et al (2002) Assessment of environmental plant stresses using multispectral steady-state fluorescence imagery, Air Pollution and Plant Biotechnology. Springer, Tokyo, pp 321–341
LandInfo: "Buying Satellite Imagery: GeoEye, WorldView 1, 2, 3, QuickBird, IKONOS, Pléiades." [Online]. Available: http://www.landinfo.com/satellite-imagery-pricing.html. Assessed on 10 Jan 2018
Liaghat S, Balasundram SK (2010) A review: the role of remote sensing in precision agriculture. Am J Agric Biol Sci 5(1):50–55
Matese A et al (2015) Intercomparison of UAV, aircraft and satellite remote sensing platforms for precision viticulture. Remote Sens 7(3):2971–2990
Nguy-Robertson A et al (2012) Green leaf area index estimation in maize and soybean: combining vegetation indices to achieve maximal sensitivity. Agron J 104(5):1336–1347
Primicerio J et al (2012) A flexible unmanned aerial vehicle for precision agriculture. Precis Agric 13(4):517–523
RAMI: Radiative Transfer Model Intercomparison (RAMI). http://rami-benchmark.jrc.ec.europa.eu/HTML/RAMI3/MODELS/4SAIL2/4SAIL2.php. Accessed on 06 Jan 2018
Rondeaux G, Steven M, Baret F (1996) Optimization of soil-adjusted vegetation indices. Remote Sens Environ 55(2):95–107
Roujean J-L, Breon F-M (1995) Estimating PAR absorbed by vegetation from bidirectional reflectance measurements. Remote Sens Environ 51(3):375–384
Satimagingcorp World view-4 Satellite imagery and satellite sensor specifications | satellite imaging corp. [Online]. Available: http://www.satimagingcorp.com/satellite-sensors/geoeye-2/. Assessed on 10 Jan 2018
Schmale III, David G, Dingus BR, Reinholtz C (2008) Development and application of an autonomous unmanned aerial vehicle for precise aerobiological sampling above agricultural fields. J Field Robot 25(3):133–147
Schneider K (2003) Assimilating remote sensing data into a land-surface process model. Int J Remote Sens 24:2959–2980
Stimson HC et al (2005) Spectral sensing of foliar water conditions in two co-occurring conifer species: Pinus edulis and Juniperus monosperma. Remote Sens Environ 96(1):108–118
Tokekar P et al (2016) Sensor planning for a symbiotic UAV and UGV system for precision agriculture. IEEE Trans Robot 32(6):1498–1511
Wallace C, Walker J, Skirvin S, Patrick-Birdwell C, Weltzin J, Raichle H (2016) Mapping presence and predicting phenological status of invasive buffelgrass in southern Arizona using MODIS, climate and citizen science observation data. Remote Sens 8(7):524
Wallach D, Makowski D, Jones JW, Brun F, Jones JW (2014) Working with dynamic crop models. Academic, Cambridge, MA, pp 407–436
Watanabe K, Guo W, Arai K, Takanashi H, Kajiya-Kanegae H, Kobayashi M, Yano K, Tokunaga T, Fujiwara T, Tsutsumi N, Iwata H (2017) High-throughput phenotyping of sorghum plant height using an unmanned aerial vehicle and its application to genomic prediction modeling. Front Plant Sci 8:421
Zarco-Tejada PJ et al (2001) Scaling-up and model inversion methods with narrowband optical indices for chlorophyll content estimation in closed forest canopies with hyperspectral data. IEEE Trans Geosci Remote Sens 39(7):1491–1507

Zarco-Tejada PJ, González-Dugo V, Berni JAJ (2012) Fluorescence, temperature and narrow-band indices acquired from a UAV platform for water stress detection using a micro-hyperspectral imager and a thermal camera. Remote Sens Environ 117:322–337

Zarco-Tejada PJ et al (2013) Estimating leaf carotenoid content in vineyards using high resolution hyperspectral imagery acquired from an unmanned aerial vehicle (UAV). Agric For Meteorol 171:281–294

Zhang C, Kovacs JM (2012) The application of small unmanned aerial systems for precision agriculture: a review. Precis Agric 13(6):693–712

Chapter 3
Unmanned Aerial Vehicle (UAV)-Based Hyperspectral Imaging System for Precision Agriculture and Forest Management

Junichi Kurihara, Tetsuro Ishida, and Yukihiro Takahashi

Abstract Hyperspectral imaging is a powerful tool for remote sensing of vegetation and environment. Although unmanned aerial vehicles (UAVs) are increasingly utilized as a new platform for remote sensing, a conventional push-broom spectrometer is not suitable for hyperspectral imaging with commercial-grade UAVs. This paper presents a hyperspectral imaging system equipped with a sequential two-dimensional spectral imager, which is more practical for UAV-based hyperspectral imaging. Liquid crystal tunable filter technology, which is also used for spaceborne imagers on microsatellites, is applied to the UAV-based hyperspectral imager for wavelength scanning in 460–780 nm. The system has a total weight of 1.5 kg, and it is designed to operate remotely with multirotor UAVs. In this paper, image processing and analysis of the acquired hyperspectral images are also described in detail with examples. This system can be widely used for UAV-based hyperspectral imaging, especially in precision agriculture and forest management.

Keywords UAV · Hyperspectral imaging · Tunable filter · Precision agriculture · Image analysis

3.1 Introduction

Hyperspectral imaging by airborne/spaceborne sensors can provide detailed information on spectral reflectance of the Earth's surface illuminated by sunlight, and it has been used for a wide variety of remote sensing applications, e.g., land cover and vegetation classification (Im and Jensen 2008; Govender et al. 2007), forest diversity assessment (Ghiyamat and Shafri 2010), coastal ocean environment (Ryan et al. 2014), and mineral mapping (van der Meer et al. 2012). The

J. Kurihara (✉) · T. Ishida · Y. Takahashi
Faculty of Science, Hokkaido University, Sapporo, Japan
e-mail: kurihara@sci.hokudai.ac.jp

R. Avtar, T. Watanabe (eds.), *Unmanned Aerial Vehicle: Applications in Agriculture and Environment*, https://doi.org/10.1007/978-3-030-27157-2_3

hyperspectral imaging sensor equipped conventionally on airborne/spaceborne platforms is a push-broom/line-scanning spectrometer, which records spectral information in a linear field of view and scans across the surface with the movement of the platform. The spatial resolution of the push-broom spectrometer is constrained by the altitude and the velocity of the platform, and thus, the typical spatial resolution is ~1 m for manned aircraft and ~30 m for satellites.

Recently, unmanned aerial vehicles (UAVs) are utilized increasingly as a new platform for remote sensing (Pajares 2015). Compared to manned aircraft and satellites, UAVs can offer higher spatial resolution to hyperspectral imaging by low-altitude and low-velocity flights. UAV-based hyperspectral imaging can contribute to precision agriculture and forest management that require high spatial resolution remote sensing from leaf to plant scales. Many UAV-compatible hyperspectral sensors have already been commercially available (Adão et al. 2017). Currently, most of the UAV-based hyperspectral sensors are a push-broom spectrometer, which is employed conventionally by the airborne hyperspectral imaging systems. However, a push-broom spectrometer is quite sensitive to the performance of the inertial measurement unit implemented on UAVs, and the use of consumer-grade UAVs for the platform of a push-bloom spectrometer will result in low-quality georeferencing and imaging (Habib et al. 2017). This limitation makes it difficult to utilize a push-broom spectrometer effectively for UAV-based hyperspectral imaging.

Two-dimensional (2D) spectral imagers, which can record 2D images of spectral bands instantaneously, are more practical for UAV-based hyperspectral imaging in terms of quality of georeferencing and imaging. Even with consumer-grade UAVs, the obtained 2D images are georeferenced easily by modern image processing techniques such as feature-based image matching. Aasen et al. (2018) categorized 2D spectral imagers for UAV-based remote sensing systems according to their technologies, e.g., multi-camera 2D imagers, sequential 2D imagers, and snapshot 2D imagers. A multi-camera 2D imager integrates several cameras of different spectral bands into a system. Several vegetation indices and biophysical parameters can be calculated from their spectral bands (Albetis et al. 2017). A sequential 2D imager uses a single camera with a tunable filter, which can change its spectral bands sequentially. Spectral features can be extracted from a sequence of spectral images after band-to-band registration, and they are investigated mostly using a machine learning technique (Näsi et al. 2018). A snapshot 2D imager can record its spectral bands at the same time. Although it has the great advantage that there is no need to conduct band-to-band registration, the number of pixels is still not enough to cover a large area with a high spatial resolution (Aasen et al. 2015).

Kurihara et al. (2018) applied liquid crystal tunable filter (LCTF) technology to a sequential 2D imager mounted on the RISING-2 microsatellite, which is a 50-kg platform developed jointly by the Tohoku and Hokkaido universities in Japan (Sakamoto et al. 2016). Following the success of the RISING-2, sequential 2D imagers with the LCTF were also mounted on subsequently launched microsatellites, e.g., the DIWATA-1, DIWATA-2, MicroDragon, and Rapid International Scientific Experiment Satellite (RISESAT). The advantage of the LCTF is that no mechanical component is used for the filter tuning. This allows the LCTF to have a compact size for easy installation, a low power consumption for wavelength scanning, and a high tolerance for vibration, shock, and vacuum environment; thus, the LCTF can be applied to a wide range of remote sensing platforms other than satellites. Consequently, a UAV-based hyperspectral imaging system with the LCTF was also developed in Hokkaido University, and it has already been used for various applications, e.g., the classification of vegetation in the field (Ishida et al. 2018).

This paper presents the specifications and the operation of the latest UAV-based hyperspectral imaging system that uses the LCTF for a sequential 2D imager. The paper also describes the image processing required to obtain high quality hyperspectral imaging dataset, which is applicable to precision agriculture and forest management.

3.2 UAV-Based Hyperspectral Imaging System

The development of the UAV-based hyperspectral imaging system in Hokkaido University commenced in 2012 as a spin-off product from the spaceborne multi-spectral sensors for microsatellites. The prototype of the system has 6.3 kg in weight including a tablet computer and a battery, and the flight test mounted on a fixed wing UAV was conducted in Indonesia in October 2012. Since then, several types of the UAV-based hyperspectral imaging systems have been developed and tested in some countries such as the Philippines (Ishida et al. 2018), and its practical utility has been improved gradually. In this section, the specification and operation of the latest UAV-based hyperspectral imaging system are described in the following subsections.

3.2.1 Specifications

The UAV-based hyperspectral imaging system consists of a sequential 2D imager, a controller, a computer, a battery, and cables (Fig. 3.1). The sequential 2D imager, which uses the LCTF for wavelength scanning and a monochrome charge-coupled device (CCD) image sensor for imaging, was designed and manufactured by Genesia Corp. (Tokyo, Japan). The LCTF is a type of optical band-pass filter that is composed of several stacked layers of liquid crystal sandwiched between crossed polarizers, and its transmission wavelength is controlled by square-wave voltages that are applied to each layer. The voltage-controlling circuit, which is connected to the LCTF of the imager, is housed separately in the controller. The approximate size and weight of the LCTF are a cube of side 30 mm and 80 g, respectively. The power consumption of the voltage-controlling circuit is 0.2 W. The central wavelength of the LCTF is electrically selectable at 1-nm intervals from 460 to 780 nm. The peak transmittance and the full width at half maximum increase with the central wavelength from 6% and 6 nm (at 460 nm) to 15% and 23 nm (at 780 nm), respectively.

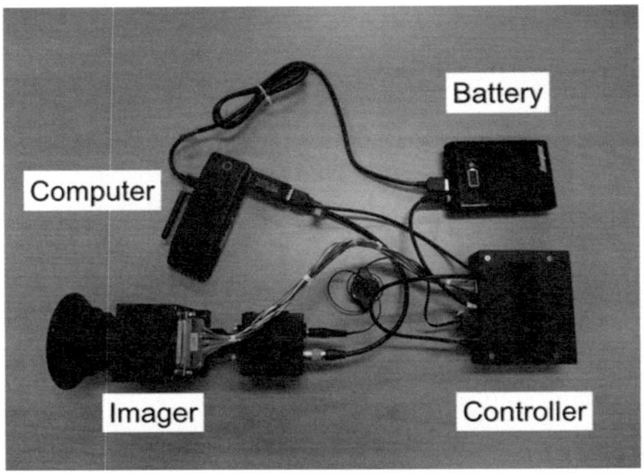

Fig. 3.1 Appearance of the UAV-based hyperspectral imaging system

As the LCTF uses the polarizers, a depolarizer is placed in front of the LCTF to scramble the polarization of incoming light. To maintain the LCTF performance over a wide temperature range of 2–38 °C, a temperature sensor is attached to the LCTF, and the optimum voltages at the measured temperature are applied automatically using a look-up table stored in the voltage-controlling circuit. The response time for switching from one central wavelength to another depends on the temperature and on the combination of wavelengths. The exposure of the CCD is designed to start after sufficient time has elapsed for the switching of the wavelengths. The computer to control the imager and the controller is a commercial stick personal computer (PC) Diginnos Stick DG-STK4D (Thirdwave Diginnos Co., Ltd., Tokyo, Japan), which uses the Windows operating system. A dedicated control software for the imager is installed on the stick PC, and it can set parameters for hyperspectral imaging, e.g., wavelengths, exposure time, gain, and number of images to be captured. The stick PC does not have a display and input devices, and it is monitored and controlled remotely from another computer via Wi-Fi during the flight. The captured images are stored on the stick PC and retrieved after the flight. The battery to provide power to the controller and the computer is a lithium-ion polymer battery Energizer XP8000A (XPAL Power Inc., Modesto, CA, USA), which is commercially available for mobile devices. Table 3.1 summarizes the specifications of the UAV-based hyperspectral imaging system.

3.2.2 Operation

Hyperspectral imaging by the sequential 2D imager is enabled by changing spectral bands sequentially. Considering the bandwidth of the LCTF, the central wavelength of the imager is changed typically at 10-nm intervals; thus, a total of 33 spectral bands are acquired sequentially in the wavelength range of 460–780 nm. The exposure time is usually adjusted in the range of 3–15 ms depending on the elevation of the sun and the cloud cover of the sky. The frame rate is normally 1–2 frames/s

Table 3.1 Specifications of the UAV-based hyperspectral imaging system

Wavelength range	460–780 nm	
Minimum wavelength interval	1 nm	
Bandwidth	6–23 nm (FWHM)	
Frame rate	1–2 frames/s	
Number of pixels	656 × 494 pixel	
Field of view	90° (diagonal)	
Operating temperature	2–38 °C	
Components	Size	Weight
Imager	220 × 90 × 90 mm	790 g
Controller	110 × 90 × 35 mm	230 g
Computer	123 × 59 × 22 mm	95 g
Battery	109 × 74 × 23 mm	225 g
Cables		160 g
Total weight		1500 g

because of the response time of the LCTF and the processing time of the image. Accordingly, it takes 25–30 s for the acquisition of an image sequence of the 33 spectral bands.

The UAV-based hyperspectral imaging system is suitable for use with multirotor UAVs, which can keep a stationary flight during the image sequence acquisition. Multirotor UAVs can also provide easier control, safer landing, and lower prices, compared to fixed wing UAVs. As the imager has a wide field of view of 90° diagonally, it can view an area of 1.6 h × 1.2 h on the ground, where h is the flight altitude of the UAV (Fig. 3.2a). In Japan, UAVs in unrestricted areas are required to stay below 150 m above ground level by the Civil Aeronautics Act. Therefore, the maximum area acquired by the imager is 240 m × 180 m = 43,200 m^2 (i.e., 4.32 ha). In order to cover the wider area, the UAV needs to move horizontally to the adjacent area and then capture the image sequence again. The overlap between the images of adjacent areas requires at least 20% of the image because of the attitude fluctuation of the UAV and the distortion of the image as described later. The effect of the attitude fluctuation can be reduced significantly by using a gimbal, which can stabilize a camera on UAVs. As shown in Fig. 3.2b, the system can be mounted on a commercially available gimbal RONIN-MX (DJI, Shenzhen, China), which is compatible with a DJI's hexacopter UAV Matrice 600 (Fig. 3.2c).

The spatial resolution of the imager is also expressed as a function of the flight altitude. The distance between pixel centers measured on the ground is approximately 2.4×10^{-3} h; thus, the spatial resolution is higher than 0.36 m in the area where the flight altitude limit is 150 m. If the attitude fluctuation of the UAV is larger than 2.4×10^{-3} rad during the exposure time of the imager, the image blurring will appear in the acquired image. Assuming that the exposure time is 10 ms, the allowable attitude fluctuation is 0.24 rad/s (i.e., 14 deg/s). When the operation of the UAV without a gimbal is carried out in windy conditions, the exposure time needs to be reduced to avoid the image blurring.

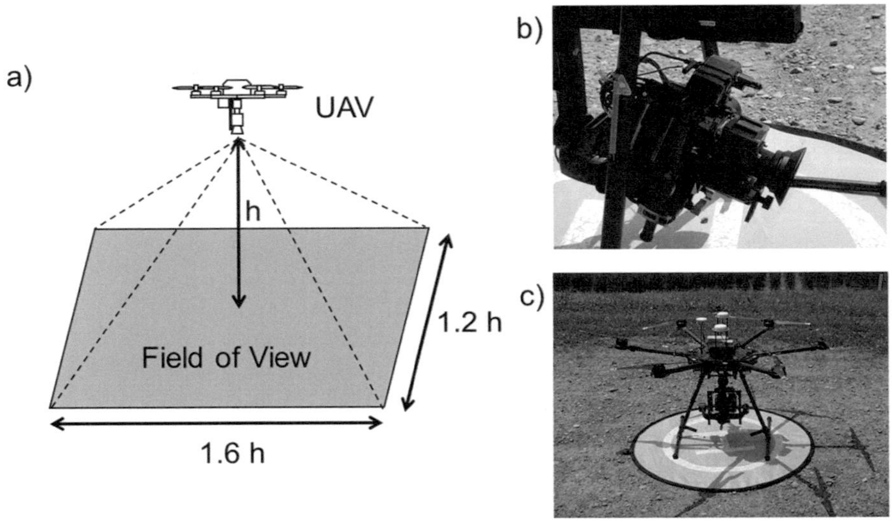

Fig. 3.2 (**a**) Schematic of the field of view of the system and (**b**) the system mounted on a gimbal RONIN-MX and (**c**) a UAV Matrice 600

3.3 Hyperspectral Image Processing

Image processing and analysis of the obtained hyperspectral images comprises five main steps: radiometric calibration, camera calibration, band-to-band registration, conversion to reflectance, and further analysis. In this section, the five main steps are described in detail in the following subsections.

3.3.1 Radiometric Calibration

Sensitivity of an image sensor cannot be completely uniform over the whole image, and hence, each pixel has a slightly different sensitivity. In addition, optical system causes vignetting, which is a reduction of the brightness in an image toward the periphery from the center. Nonuniformity in an image related to these intrinsic factors of the imager can be measured using a uniform light source in a laboratory experiment. Figure 3.3a shows a uniform light source HELIOS USLR-D12L-NMNN (Labsphere, Inc., North Sutton, NH, USA) used for the radiometric calibration. The uniform light source employs an integrating sphere whose inside is coated by a highly diffuse reflecting material, Spectralon. The spectral radiance of the integrating sphere is calibrated with equipment and methods traceable to the US National Institute of Standards and Technology.

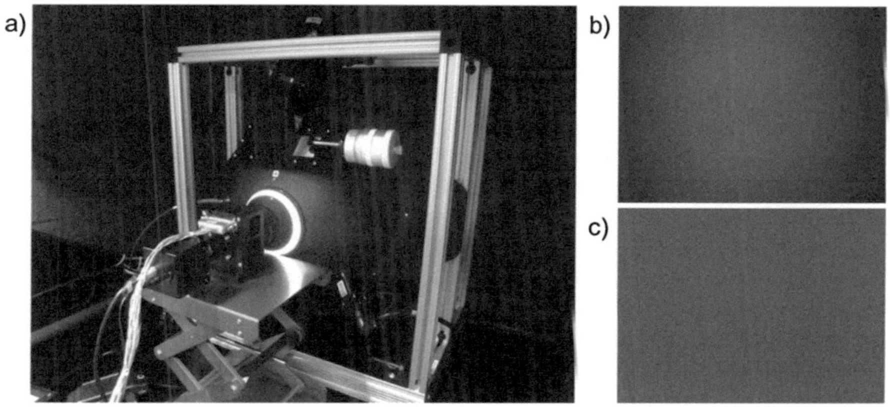

Fig. 3.3 (**a**) Photograph of the radiometric calibration with a uniform light source, (**b**) the acquired image, and (**c**) the calibrated image

In the laboratory experiment, digital numbers of the pixels in the hyperspectral images taken by the imager are related to the spectral radiance of the integrating sphere as

$$L(\lambda) = C_i(\lambda) \times (DN_i(\lambda) - B)/T_{\exp}, \tag{3.1}$$

where $L(\lambda)$ is the spectral radiance of the integrating sphere at the wavelength λ, $C_i(\lambda)$ is the unit conversion coefficient of the ith pixel at the wavelength λ, $DN_i(\lambda)$ is the digital number (i.e., brightness value) of the ith pixel at the wavelength λ, B is the offset of the pixels, and $T_{\exp}$ is the exposure time spent on acquisition of the image. By using the unit conversion coefficients derived from Eq. 3.1, digital numbers of the hyperspectral images taken from the UAV can be converted to the spectral radiance as

$$L_i(\lambda) = C_i(\lambda) \times (DN_i(\lambda) - B)/T_{\exp}, \tag{3.2}$$

where $L_i(\lambda)$ is the spectral radiance of the ith pixel at the wavelength λ. Figure 3.3b shows the image of the inside of the integrating sphere taken by the imager at 600 nm, and Fig. 3.3c shows the image calibrated by the measured unit conversion coefficients. As can be seen, nonuniformity in the original image, whose brightness is lower in the periphery than in the center, is indistinctive in the calibrated image. As the unit conversion coefficients vary slightly from one imager to another, the nonuniformity measurement for the radiometric calibration is necessary for all the imagers at least once.

3.3.2 Camera Calibration

The wide field of view of the imager causes significant barrel distortion of the image. Removing the distortion is necessary for subsequent image registration. Although there are many methods to remove distortion from an image, camera calibration algorithms implemented in an open-source library, OpenCV (http://opencv.org/), were used in this study. This camera calibration is conducted by taking images of a calibration object from various angles and obtaining geometric relationship between the points on the object and the pixels in the acquired image. A black-white chessboard pattern was used for the calibration object, and the corners of squares on the chessboard pattern were detected automatically as reference points. As the size of the chessboard pattern should be large enough to cover the wide field of view of the imager, the chessboard pattern was shown on a large screen display instead of instead of being printed on a paper (Fig. 3.4a). After the camera calibration, the barrel distortion of the chessboard pattern in the original image (Fig. 3.4b) is removed entirely in the calibrated image (Fig. 3.4c). The corners of the original image are changed to acute angles and extended to the outside of the calibrated image frame, and thus, the field of view of the calibrated mage is reduced by approximately 10% from the original.

3.3.3 Band-to-Band Registration

The acquired images deviate from one another spatially due to attitude fluctuations of the UAV during the sequential image acquisition. Therefore, band-to-band registration using the feature-based matching approach is applied to a sequence

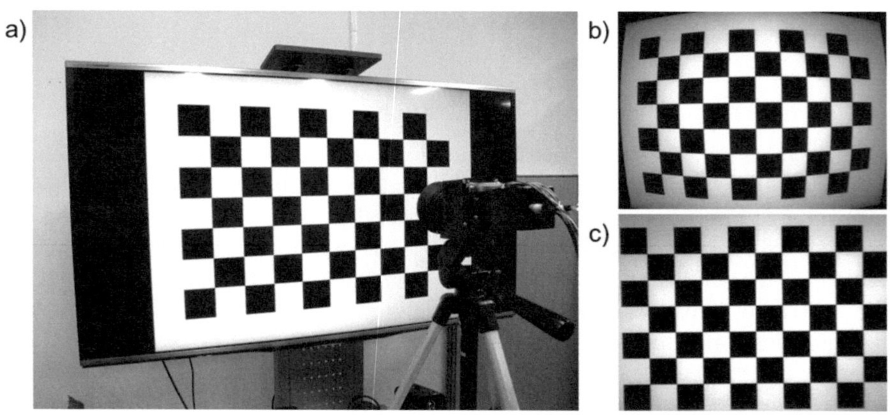

Fig. 3.4 (**a**) Photograph of the camera calibration with a displayed chessboard, (**b**) the acquired image, and (**c**) the calibrated image

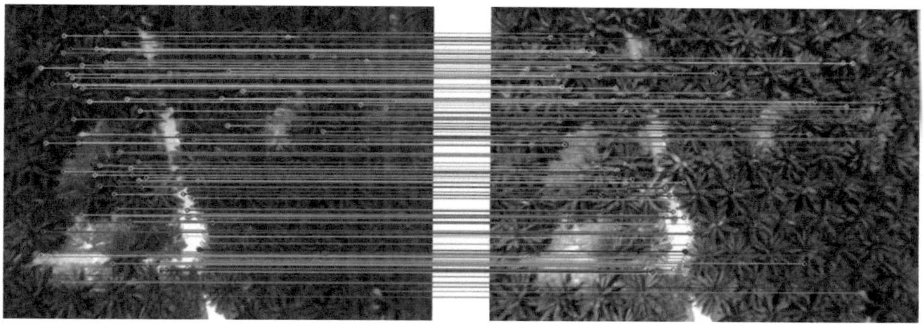

Fig. 3.5 Result of the feature matching between the images at 690 nm (left) and 700 nm (right)

of the images prior to the further image processing. Although many feature-based matching algorithms are also available in the OpenCV library, Accelerated-KAZE (A-KAZE) feature detection and description method (Alcantarilla et al. 2013) is used in this study.

In the previous study, the scale-invariant feature transform (SIFT) method (Lowe 2004), which is recognized as one of the most precise methods of feature matching, was employed. However, as the result of a comparison of the two methods, the performance of A-KAZE is at least ten times faster in computing and similar in precision compared to that of SIFT. In addition, the SIFT algorithm is patented, and a license is required for the commercial application of the algorithm.

As an example of the band-to-band registration, Fig. 3.5 shows a result of the feature matching between the spectrally sequential images at 690 nm and 700 nm. The matched feature points, which are connected and highlighted by colored lines in Fig. 3.5, can provide a homography matrix that describes a transformation from the coordinate system of the image at 690 nm to that at 700 nm. The homography matrixes are obtained for all the sequential images in the same manner, and finally, all the images are transformed into the coordinate system of the one image.

The band-to-band registration allows production of a hyperspectral cube, which is the three-dimensional dataset of hyperspectral images. In this step, the hyperspectral cube is still a radiance-based, and it is converted to a reflectance-based in the next step.

3.3.4 Conversion to Reflectance

In the field of remote sensing, hyperspectral images are analyzed generally based on the spectral reflectance, because the spectral radiance depends on time-varying spectral irradiance by sunlight. In addition, UAV-based hyperspectral imaging is available under cloudy condition, in which the spectral irradiance changes locally and temporarily. However, simultaneous measurement of the spectral radiance and the spectral irradiance requires an additional spectrometer.

If there is an object of known spectral reflectance in an image, it can be used as a reference for the other objects. Hence, a reflectance standard panel is used for the reference of spectral reflectance. Spectral reflectance of an object on the ground is defined by

$$R(\lambda) = \pi L(\lambda)/E(\lambda), \tag{3.3}$$

where $R(\lambda)$ is the spectral reflectance of the object at the wavelength λ, $L(\lambda)$ is the spectral radiance of the object measured by the imager at the wavelength λ, and $E(\lambda)$ is the spectral irradiance of the object at the wavelength λ. If the reflectance standard panel is measured simultaneously by the imager, its spectral reflectance is

$$R_s(\lambda) = \pi L_s(\lambda)/E(\lambda), \tag{3.4}$$

where $R_s(\lambda)$ and $L_s(\lambda)$ are the spectral reflectance and the spectral radiance, respectively, of the reflectance standard panel at the wavelength λ. Accordingly, the spectral reflectance of the object becomes independent of the spectral irradiance as

$$R(\lambda) = R_s(\lambda)\, L(\lambda)/L_s(\lambda). \tag{3.5}$$

By means of this equation, the radiance-based hyperspectral cube is converted to reflectance-based hyperspectral cube.

Although the reflectance standard panel is commercially available, the product that is large enough to be measured by the imager from the flight altitude of the UAV is expensive and heavy. In this study, an ethylene-vinyl acetate (EVA) foam sheet, which is marketed globally as an EVA joint mat, is adopted as a substitute for the reflectance standard panel. The advantages of the EVA mat are that it is cheap, light, and easy to spread out by jointing. Figure 3.6 shows spectral reflectance of two types

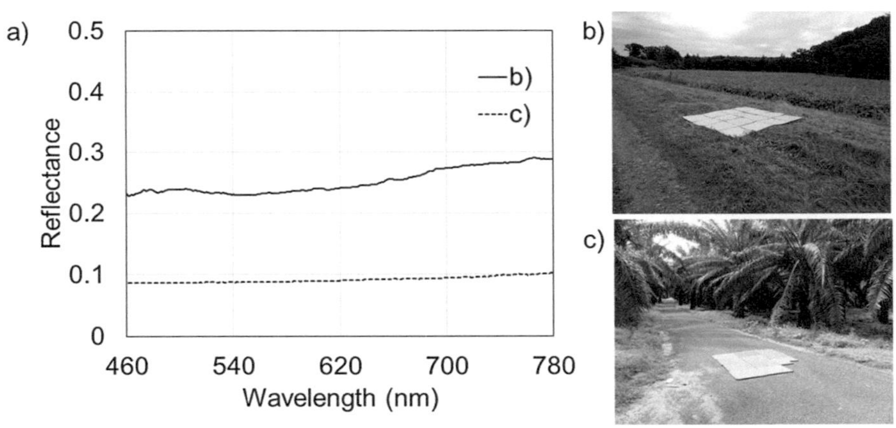

Fig. 3.6 (**a**) Spectral reflectance of the EVA mats for (**b**) and (**c**)

of the EVA mat measured by a portable spectroradiometer ASD FieldSpec 4 (Analytical Spectral Devices, Inc., Longmont, CO, USA). While the two EVA mats have quite different reflectance, their spectra are smooth enough to be used for the reference in the wavelength range of the imager.

3.3.5 Further Analysis

Further analysis of the hyperspectral cube varies with individual applications of UAV-based hyperspectral imaging. Nevertheless, there are essential information required commonly for many applications. For example, a color image is quite useful for the georeferencing of a hyperspectral image. In fact, many UAV-based hyperspectral imaging systems are equipped with a red-green-blue (RGB) camera only for that purpose (Aasen et al. 2018). Figure 3.7a shows a true color composite image produced from the hyperspectral cube. The spectral reflectance at 460–500 nm, 510–590 nm, and 600–690 nm is averaged and then assigned to the blue, green, and red channels, respectively, of the color composite image. Even though an additional weight is not given to its channels, the composite image has well-balanced natural colors. This composite image production allows the UAV-based hyperspectral imaging system to remove another RGB camera.

Another example is the normalized difference vegetation index (NDVI) image. The NDVI is widely used in multispectral remote sensing, and it is derived from the expression,

$$\mathrm{NDVI} = (\rho_{\mathrm{NIR}} - \rho_{\mathrm{RED}})/(\rho_{\mathrm{NIR}} + \rho_{\mathrm{RED}}), \tag{3.6}$$

where ρ_{RED} and ρ_{NIR} are the reflectance of red and near-infrared (NIR) bands, respectively. The NDVI is useful not only for qualitative and quantitative analysis of vegetation in multispectral imaging but also for extraction of vegetation in preprocessing of hyperspectral imaging. Figure 3.7b shows the NDVI image produced from the same hyperspectral cube with Fig. 3.7a by assigning the spectral

Fig 3.7 (**a**) A true color composite image and (**b**) the NDVI image

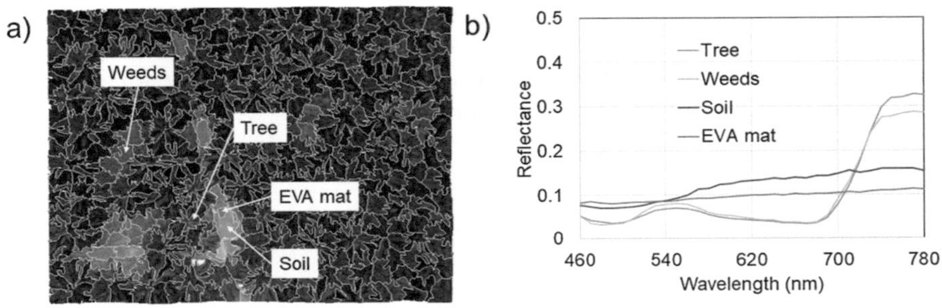

Fig. 3.8 (**a**) Superpixel segmentation of the image and (**b**) the spectral reflectance of typical objects

reflectance of 680 and 770 nm to ρ_{RED} and ρ_{NIR}, respectively, in Eq. 3.6. As seen in Fig. 3.7b, vegetation is obviously distinguishable from other objects such as soil, according to the value of the NDVI. Note that the vegetation in the image includes trees and weeds without any distinction.

In precision agriculture and forest management, hyperspectral imaging is applied typically to classification of vegetation. As hyperspectral imaging from manned aircraft and satellites presents relatively low spatial resolution images, different species of vegetation and other objects are mixed with each other in a single pixel of the acquired image. Thus, hyperspectral unmixing techniques are frequently required for the classification (Bioucas-Dias et al. 2012). On the other hand, UAV-based hyperspectral imaging can provide higher spatial resolution, and objects in the acquired image are separated clearly from each other except for their boundary pixels. Therefore, object-based classification is a more effective approach to UAV-based hyperspectral image analysis than pixel-based classification (Cao et al. 2018). Image segmentation, in which neighboring pixels are grouped into objects based on their spectral information, is the first process of object-based classification. Figure 3.8a shows an example of the image segmentation using the Simple Linear Iterative Clustering (SLIC) superpixel segmentation algorithm (Achanta et al. 2012) implemented in OpenCV. The SLIC is a simple and fast algorithm that adapts a k-means clustering approach to generate superpixels efficiently. According to their colors, neighboring pixels in the color image (Fig. 3.7a) are grouped into superpixels, which are divided by the white boundaries in Fig. 3.8a. The spectral reflectance is averaged over each superpixel (Fig. 3.8b), and they can be investigated further by using machine learning techniques.

3.4 Conclusions

UAV-based hyperspectral imaging is an emerging technology that evolves rapidly in response to the recent development of UAV and remote sensing technologies. Although future applications of UAV-based hyperspectral imaging in precision agriculture and forest management are promising, the current technologies are still under the validation phase prior to the practical use. Regarding the use of consumer-grade UAVs, a sequential 2D imager has the advantage in quality of georeferencing and imaging over a push-bloom spectrometer. In this paper, the UAV-based hyperspectral imaging system using the sequential 2D imager is described in detail. The system employs the LCTF for wavelength scanning of the sequential 2D imager, and the specifications of the system are optimized to mount on multirotor UAVs. The operation of the system can be adapted flexibly to the requirements of coverage and spatial resolution and the conditions of cloud and wind.

The high-quality dataset of UAV-based hyperspectral imaging is also required for the practical applications in precision agriculture and forest management. Image processing and analysis of hyperspectral images can be performed readily by open-source libraries and commercial software now. In this paper, the five steps of image processing of the acquired hyperspectral images are also described. The precise radiometric calibration and camera calibration are conducted to improve the quality of the acquired images based on the laboratory measurements. The accurate band-to-band registration and conversion to reflectance are important to produce the reflectance-based hyperspectral cube. Finally, some common processes prior to the further analysis are introduced with the examples. Although the real-time onboard processing is almost impossible for the UAV-based hyperspectral imaging system because of the huge amount of data, the processing time can be reduced significantly by the automated and pipelined image processing.

References

Aasen H, Burkart A, Bolten A, Bareth G (2015) Generating 3d hyperspectral information with lightweight UAV snapshot cameras for vegetation monitoring: from camera calibration to quality assurance. ISPRS J Photogramm Remote Sens 108:245–259. https://doi.org/10.1016/j.isprsjprs.2015.08.002

Aasen H, Honkavaara E, Lucieer A, Zarco-Tejada PJ (2018) Quantitative remote sensing at ultra-high resolution with UAV spectroscopy: a review of sensor technology, measurement procedures, and data correction workflows. Remote Sens 10:1091. https://doi.org/10.3390/rs10071091

Achanta R, Shaji A, Smith K, Lucchi A, Fua P, Süsstrunk S (2012) SLIC superpixels compared to state-of-the-art superpixel methods. IEEE Trans Pattern Anal Mach Intell 34:2274–2281. https://doi.org/10.1109/TPAMI.2012.120

Adão T, Hruška J, Pádua L, Bessa J, Peres E, Morais R, Sousa J (2017) Hyperspectral imaging: a review on UAV-based sensors, data processing and applications for agriculture and forestry. Remote Sens 9:1110. https://doi.org/10.3390/rs9111110

Albetis J, Duthoit S, Guttler F, Jacquin A, Goulard M, Poilvé H, Féret J-B, Dedieu G (2017) Detection of Flavescence dorée grapevine disease using unmanned aerial vehicle (UAV) multispectral imagery. Remote Sens 9:308. https://doi.org/10.3390/rs9040308

Alcantarilla PF, Nuevo J, Bartoli A (2013) Fast explicit diffusion for accelerated features in nonlinear scale spaces. Trans Pattern Anal Mach Intell 34:1281–1298. https://doi.org/10.5244/C.27.13

Bioucas-Dias J, Plaza A, Dobigeon N, Parente M, Du Q, Gader P, Chanussot J (2012) Hyperspectral unmixing overview: geometrical, statistical, and sparse regression-based approaches. IEEE J Sel Top Appl Earth Obs Remote Sens 5:354–379. https://doi.org/10.1109/JSTARS.2012.2194696

Cao J, Leng W, Liu K, Liu L, He Z, Zhu Y (2018) Object-based mangrove species classification using unmanned aerial vehicle hyperspectral images and digital surface models. Remote Sens 10:89. https://doi.org/10.3390/rs10010089

Ghiyamat A, Shafri H (2010) A review on hyperspectral remote sensing for homogeneous and heterogeneous forest biodiversity assessment. Int J Remote Sens 31:1837–1856. https://doi.org/10.1080/01431160902926681

Govender M, Chetty K, Bulcock H (2007) A review of hyperspectral remote sensing and its application in vegetation and water resource studies. Water SA 33:145–152. https://doi.org/10.4314/wsa.v33i2.49049

Habib A, Xiong W, He F, Yang HL, Crawford M (2017) Improving orthorectification of UAV-based push-broom scanner imagery using derived orthophotos from frame cameras. IEEE J Sel Top Appl Earth Obs Remote Sens 10:262–276. https://doi.org/10.1109/JSTARS.2016.2520929

Im J, Jensen J (2008) Hyperspectral remote sensing of vegetation. Geogr Compass 2:1943–1961. https://doi.org/10.1111/j.1749-8198.2008.00182.x

Ishida T, Kurihara J, Viray FA, Namuco SB, Paringit EC, Perez GJ, Takahashi Y, Marciano JJ Jr (2018) A novel approach for vegetation classification using UAV-based hyperspectral imaging. Comput Electron Agric 144:80–85. https://doi.org/10.1016/j.compag.2017.11.027

Kurihara J, Takahashi Y, Sakamoto Y, Kuwahara T, Yoshida K (2018) HPT: a high spatial resolution multispectral sensor for microsatellite remote sensing. Sensors 18:619. https://doi.org/10.3390/s18020619

Lowe DG (2004) Distinctive image features from scale-invariant keypoints. Int J Comput Vis 60:91–110. https://doi.org/10.1023/B:VISI.0000029664.99615.94

Näsi R, Viljanen N, Kaivosoja J, Alhonoja K, Hakala T, Markelin L, Honkavaara E (2018) Estimating biomass and nitrogen amount of barley and grass using UAV and aircraft based spectral and photogrammetric 3D features. Remote Sens 10:1082. https://doi.org/10.3390/rs10071082

Pajares G (2015) Overview and current status of remote sensing applications based on unmanned aerial vehicles (UAVs). Photogramm Eng Remote Sens 81:281–330. https://doi.org/10.14358/PERS.81.4.281

Ryan JP, Davis CO, Tufillaro NB, Kudela RM, Gao B-C (2014) Application of the Hyperspectral imager for the Coastal Ocean to phytoplankton ecology studies in Monterey Bay, CA, USA. Remote Sens 6:1007–1025. https://doi.org/10.3390/rs6021007

Sakamoto Y, Sugimura N, Fukuda K, Kuwahara T, Yoshida K, Kurihara J, Fukuhara T, Takahashi Y (2016) Development and flight results of microsatellite bus system for RISING-2. Trans JSASS Aerosp Technol Jpn 14:Pf_89–Pf_96. https://doi.org/10.2322/tastj.14.Pf_89

van der Meer FD, van der Werff HM, van Ruitenbeek FJ, Hecker CA, Bakker WH, Noomen MF, van der Meijde M, Carranza EJM, de Smeth JB, Woldai T (2012) Multi- and hyperspectral geologic remote sensing: a review. Int J Appl Earth Obs Geoinf 14:112–128. https://doi.org/10.1016/j.jag.2011.08.002

Chapter 4
Unmanned Aerial Vehicle (UAV) for Fertilizer Management in Grassland of Hokkaido, Japan

Kanichiro Matsumura

Abstract The use of unmanned aerial vehicles (herein after UAVs) as a tool in forage crop management is discussed. Overfertilizing has negative effects on water quality and affects health of animals and human beings, who use it. The author conducted experiments aimed at reducing fertilizer consumption. The analysis of two different images taken before and after the harvest period clearly shows a highlighted portion of the test area that might be overfertilized. UAVs equipped with cameras using both the RGB portion of the visible light spectrum and the near infrared portion of the electromagnetic spectrum (780–2500 nm) take Blue Normalized Difference Vegetation Index (BNDVI: blue band is near infrared portion in this research) images. Comparing the images can detect possible overfertilized areas. This data can then be used to adjust future fertilizer application rates. Experiments conducted in 2017 and 2018 and upcoming experiments in 2019 are discussed. Adjusting the analysis for BNDVI intensity and comparing UAV sourced images with satellite remote sensing data began in 2018.

Keywords UAVs · BNDVI · Overfertilized

4.1 Introduction

Technological advances have led to cost-effective UAVFs becoming widely used in both the civil engineering and agricultural fields. UAVs are now expected to become useful tools for farmers. They can collect area data from remote inaccessible regions but can also quickly and inexpensively collect data from more accessible areas with significant savings in time. Demographic changes in eastern Hokkaido have resulted in labor shortages so agriculture practitioners require more efficient technologies.

K. Matsumura (✉)
Tokyo University of Agriculture, Abashiri, Hokkaido, Japan
e-mail: km205693@nodai.ac.jp

R. Avtar, T. Watanabe (eds.), *Unmanned Aerial Vehicle: Applications in Agriculture and Environment*, https://doi.org/10.1007/978-3-030-27157-2_4

Cattle should not consume overfertilized forage which can lead to health problems. Environmentally overfertilization can lead to contamination of both surface runoff and groundwater. The author analyzed spatial growth patterns in grassland with the aim that the data could be used to both optimize fertilizer use and reduce its overuse. A series of pictures with coordinates can be transformed into a three-dimensional picture. Structure from motion (SfM) technology and UAV technology can build a 3-D model from pictures (Inoue et al. 2014). The author surveyed farmers, and farmers know the effects of fertilizer are influenced by variables including land surface topography and soil strata. UAVs can quickly produce a digital elevation model (DEM). Satellite image data can be used for analyzing crop production. In this study, the author started to use satellite data provided by Planet (Planet company 2018) for grassland management studies from a distribution provider (Space Agri 2018). Planet can provide a picture for every place on earth within a 3-meter resolution, and this can be used commercially in agriculture. Planet can provide a picture every day. Satellite images are influenced by cloud cover. During June 2018, clouds prevented the author from getting surface information for the grassland studies. UAVs have an advantage to get the information under clouds.

4.2 Materials and BNDVI

A DJI-manufactured Phantom 3 Professional UAV was primarily used for this research. The author also examined the possibility of using a commercial fixed wing UAV (Fig. 4.1).

The author had collaborated with a Malaysian scientist who had custom-built fixed wing UAVs for studying palm tree plantations. Custom-built UAVs have advantages in increased payload and can also be designed with special features. It

Fig. 4.1 A Phantom 3 and a fixed wing UAV (left) and the autopilot system for the fixed wing UAV (right)

Fig. 4.2 A flight plan plotted using Mission Planner (left) and the flight plan shown when viewed with the UAV Litchi software (right)

is favorable for UAVs to fly preprogrammed paths, thus making shooting conditions almost completely similar. The Autopilot software provides this feature. Mission Planner ™ and Litchi™ software are used in this regard. The flight plan is generated using Mission Planner and uploaded to the Mission Hub-Litchi digital cloud (Fig. 4.2). Mission Hub-Litchi works as the communications link between Mission Planner and Litchi.

The Normalized Difference Vegetation Index (NDVI) is widely used for studying the intensity of plant activity. Green plant leaves absorb the RGB color portion of the light spectrum ranging from 0.4 to 0.7 μm and reflect light from the infrared band of the spectrum ranging from 0.7 to 1.3 μm. Healthy vegetation (chlorophyll) reflects more near-infrared and green light compared to other wavelengths and absorbs more red and blue parts of the light spectrum (GIS Geography 2019) NDVI changes according to both growth stages and growing conditions (Lu et al. 2015). This research used the Blue Normalized Difference Vegetation Index (BNDVI) instead of the NDVI for the analysis. The infrared lens “NDVI7” was obtained from the USA (IR-Pro 2017). The image is filtered with the NIR, green, and blue channels. The formula is expressed below:

$$\text{BNDVI} = (\text{NIR} - \text{Blue})/(\text{NIR} + \text{Blue}) \tag{4.1}$$

Data images in 2017 and 2018 were mainly collected from a rotary platform UAV equipped with a near-infrared camera. The author intends to use a multispectral sensor camera (Sequoia ™ 2019) which can detect the red-edge band portion of the spectrum. The red-edge bands particularly illustrate the changes in the reflectance between the red portion of the spectrum and the near-infrared band. An examination of the reflectance graph shows a sharp change in the reflectance values from the red to infrared band. The missing portion of the reflectance therefore is the red-edge band. This band shows information about the changes in the phenological properties of vegetation as well as showing species level changes.

Fig. 4.3 The BNDVI difference subtracting Jun 14th from Jul 13th (left), BNDVI difference subtracting Jul 13th from Aug 17th (middle), and a composite image combining both the left and middle images

4.3 Experiments in 2017

Adjusting fertilizer levels is an urgent matter both due to environmental impact and budgetary costs. The author developed a method to identify overfertilized areas (Matsumura and Inoue 2018). An analysis of two images taken before and after the harvest period shows that the highlighted portion of the test area might be overfertilized. Citation shows that this difference is indeed weaker or stronger growth and is due to fertilizer levels. The grassland is cut three times a year. In 2017 the first cutting was June 17th, the second August 3rd, and the third cutting on Oct 14th. BNDVI images from May 17th, Jun 14th, Jul 13th, and Aug. 17th to Sep 19th were selected for analysis (Fig. 4.3). The harvested area shows lower BNDVI values. The BDVI values from before the harvest were subtracted from the BNDVI values after harvest. To highlight the differences if the subtracted value was over (less than) "0," it was colored in black (white). The Jun 14th image is subtracted from the July 13th image (left in Fig. 4.3) On Jun 17th, the first harvest was conducted. The Jul 13th image was subtracted from the August 17th image (middle in Fig. 4.3). On Aug 3rd, the second harvest was conducted.

After harvesting, there is always a decrease in BNDVI values; however there are some areas that show higher values. The analysis shows the highlighted areas that might be overfertilized (right in Fig. 4.3). Intensity of sunshine changes and adjustment of BNDVI values mentioned following chapter is not conducted in 2017.

4.4 Experiments in 2018

The 2017 experiments highlight possible overfertilized area. In 2018, the first cutting was conducted on Jun 18th, the second between Aug 28th and 29th, and the third on Sep 19th. The before/after paired images were obtained from Jun 11th and 21st, Aug 27th and Aug 30th, and Sep 18th and Sep 20th. Analysis of the images included an adjustment using white circle shape board (herein after, white reflector). White

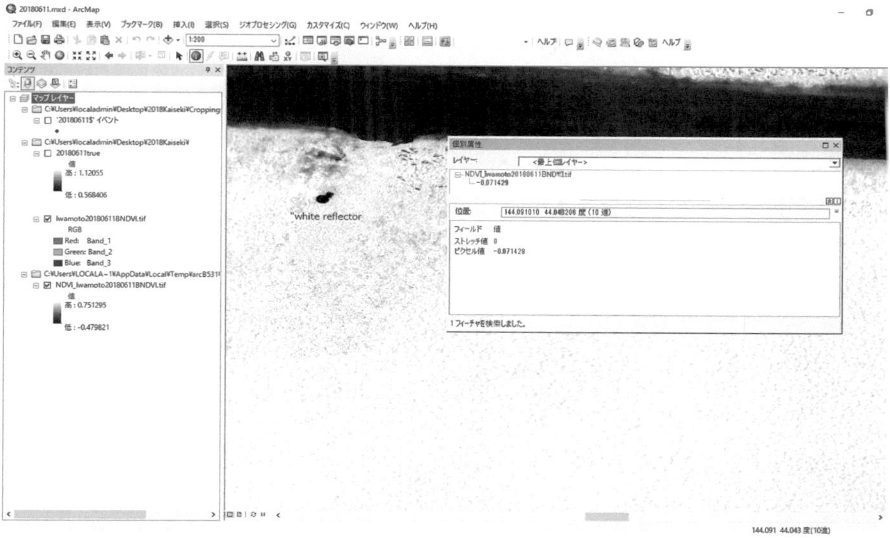

Fig. 4.4 The ortho-rectified BNDVI image from the ArcGIS software after adjusting for the "white reflector"

reflector reflects sunlight and the BNDVI values are expected to be zero. The difference between the expected value (=zero) and the observed value was adjusted.

The following illustrates the series of data corrections. In Fig. 4.4 the white reflector calibration method was applied to the Jun 11th image. The white reflector value was −0.064718. The adjustment adds 0.064718 in value to the whole BNDVI image using the raster function that accompanies the ArcGIS software. A GreenSeeker™ handheld crop sensor was used for the second stage of the adjustments (Fig. 4.5). The GreenSeeker, a hand tool, claims to be designed to help create a high-yielding environment while reducing nutrient input costs and eliminating the need for application maps. It can be operated at all times and in all weathers (GreenSeeker handheld crop sensor 2019). The author considered the NDVI values obtained from the sensor (in situ) to be the true value. In situ NDVI values were collected using the sensor together with the UAVs onboard camera. The handheld sensor itself is not equipped with GPS positioning software. The author compared the sensor collected NDVI data with the adjusted UAV collected data. The adjustment was calibrated by using a ratio of data from the handheld sensor and BNDVI UAV sourced data. If there was more than one reading, the average value was used for calibration. In the following case, two readings were averaged. The formula is as follows:

$$\left(\text{``NDVI\_Iwamoto20180611BNDVI.tif} + 0.064718\right) \\ * \left((0.77 + 0.79)/2\right) \left((0.657754 + 0.538462)/2 + 0.064718\right)$$

where Iwamoto is the name of the study area and user of the forage crop, NDVI_Iwamoto20180611BNDVI.tif is the ortho-image by UAVs,

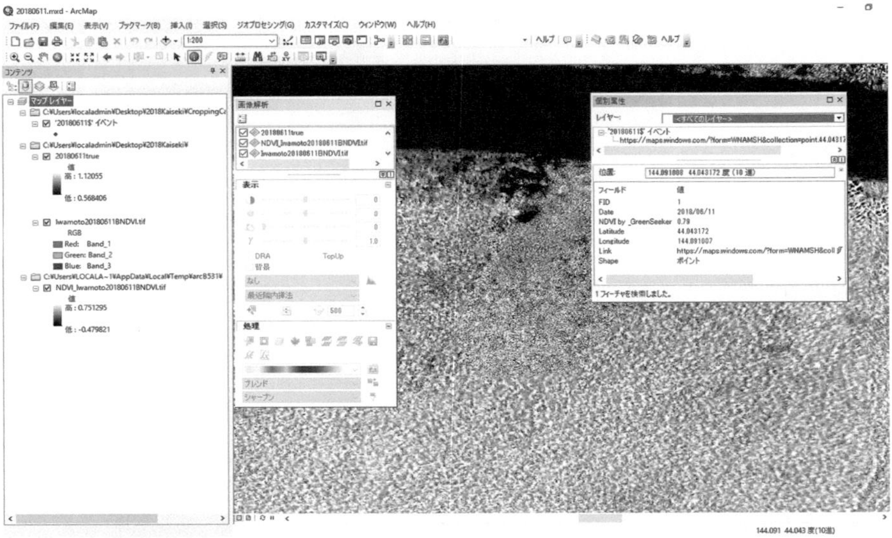

Fig. 4.5 In situ NDVI values taken with the GreenSeeker sensor adjusted for the values taken with the UAVs onboard camera

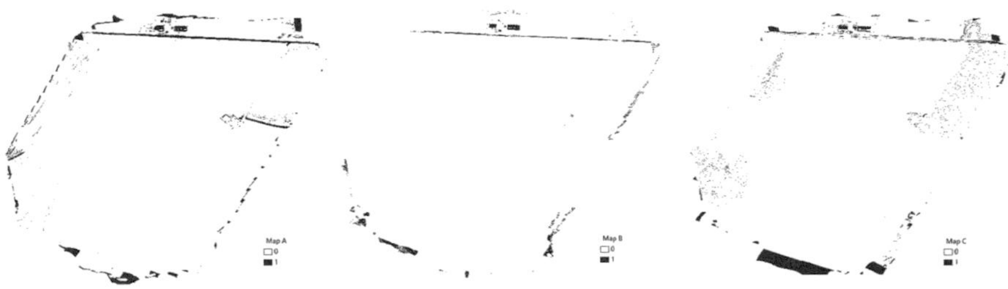

Fig. 4.6 Subtracted Map A (left), Map B (middle), and Map C (right)

(0.77 + 0.79)/2 is the average of two BNDVI readings obtained from the ortho-UAV image,

(0.647754 + 0.538462)/2 is the average of two sensor readings corresponding with the UAV position, and

0.064718 is the adjustment value calculated using the white reflector.

The same procedure was followed for June 21st (white circle value −0.053996, GreenSeeker value 0.3), July 27th (white circle value −0.121951, GreenSeeker value 0.77), July 30th (white circle value −0.086207, GreenSeeker value 0.77 and 0.79), September 18th (white circle value −0.052411, GreenSeeker value 0.8), and September 20th (white circle value −0.064718, GreenSeeker value 0.43). Highlighting the differences, if the adjusted calculated value was greater than "0," it is colored black, and if less than "0," it is colored white. The June 11th image was subtracted from the June 21st image and is defined as Map A in Fig. 4.6 (first harvest June 18th). July 27th image is subtracted from July 30th image and defined as Map B. Between

Fig. 4.7 Subtracted Map D (left), Map E (middle), and multiplied Map F (right)

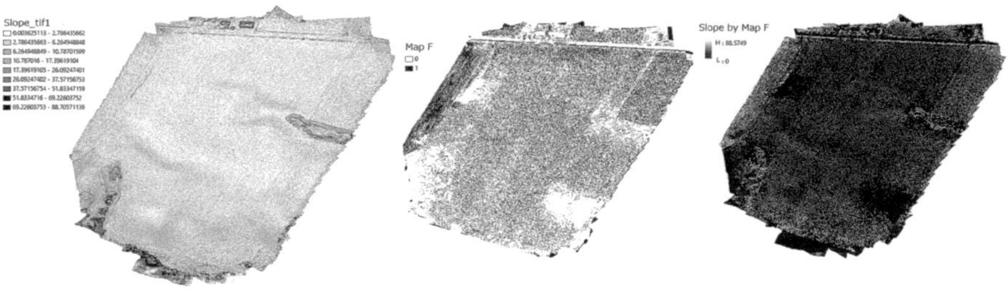

Fig. 4.8 Slope data (left), might be less fertilized area (middle), and multiplied map (right)

July 27th and 30th, second harvesting is conducted. September 18th image is subtracted from September 20th image and defined as Map C. Each of Map A, B, and C shows that before harvesting shows higher value than after harvesting.

To highlight the difference of BNDVI intensity during season, June 11th image is subtracted from July 27th image and defined as Map D. July 27th image shows higher BNDVI values than that of June 11th. July 27th image is subtracted from September 18th image and defined as Map E. September 18th image shows higher BNDVI values than that of July 27th. Multiplying Map D by Map E is defined as Map F. Among Map F, those area that show zero might be more fertilized than usual (Fig. 4.7). According to farmer's survey, while fertilizing, topography is taken into account, and top of topography, less fertilizing is taken into account.

Topography is discussed. Slope map is generated from digital elevation map and it is multiplied with Map F. Those area in Map F showing "0" might be more fertilized, or those area showing "1" might be less fertilized area (Fig. 4.8). Those area showing steep slope correspond to less fertilized area.

4.5 Remote Sensing Data

The combination of remote sensing data and UAVs can be a complimentary relationship. Many companies entering space-related business have been established across the world. In 2015, the total investments in space venture companies in the world are increasing, reaching $ 800 million per year. There has been a movement to build real-time earth observation networks circling several tens to over 100 satellites around the earth since 2016. Planet Labs Inc. in the USA is implementing acquisition of the project of Terra Bella including image satellites of SkySat Earth from Google to provide satellite imagery to Google Maps, etc. (Table 4.1). Planet Labs launched 88 of the 3 m resolution satellites using an Indian rocket on February 14, 2017 and successfully put them in their orbits. On July 14, 2017, it also launched 48 satellites on a Russian rocket, and it has 152 satellites orbiting around the earth when already-launched operation equipment is also included.

Planet data can shoot 3-m resolution image every day if it were fine.

1.0-m resolution image: The kind of building can be confirmed. Cars are also recognized.
2.5-m resolution image: The kind of building can somehow be confirmed. Although the existence of objects can be recognized, it is not known that it is a car.
5.0 m resolution image: Although the existence of a building can be recognized, it is not known what kind of building it is. Cars cannot be detected.
10.0-m resolution image: Large buildings can somehow be recognized. Cars cannot be detected.

Images before and after harvesting were selected. From experiments in 2018, through the kind support of Space-Agri company founded by an engineer, Mr. Takashi Seshimo, the author has started to obtain Planet data for study area from 2018. During season, harvesting is conducted three times, and two images crossing harvest period are selected to compare with UAV images. In 2018, harvesting is conducted in June (Fig. 4.9), July (Fig. 4.10), and September (Fig. 4.11).

In each harvesting period, subtracting "after image" from "before image" is conducted. Comparing first harvesting time difference with second harvesting time difference shows that most of the area shows higher NDVI value. Comparing second harvesting time difference with third harvesting time difference shows different NDVI values and highlights less fertilized area. This area corresponds partly to steep slope portion (Fig. 4.12).

Satellite data can be viewed the following day. The 3-m resolution makes it possible to understand the situation of fields. Research to create wheat growth map is already in practical stage in combination with fertilization machinery. Sometimes the data does not come out, or sometimes it cannot be seen in the following day but can be only viewed several days later. The problem is caused by light cloudiness. During 2018 season, in June, cloudy days make it impossible for users to obtain image. However, Space-Agri company provides incredibly low cost such as less than 2 dollars/10a/year. A total of 107 farmers and 292 organizations have monitored the efforts at the end of October since April 2017, and approx. 45,000 farms and 114,000 ha have been registered.

Table 4.1 Comparison of remote sensing data (Provided by Space-Agri company)

	Latest satellite remote sensing	Conventional satellite remote sensing	References	
	Planet, etc.	RapidEye, etc.	Drone	Crop Spec[a]
Shooting method	Optical artificial satellite for which 120–150 satellites were arranged over the earth	Optical artificial satellite for which five satellites were arranged on the same orbit	Multicopter- and airplane-type flying objects	Device loaded on the roof of a tractor
Resolution	3 m	6 m	0.3 m	
Observation width	24 km	77 km	100 m	10 m
Frequency	All sunny days	Once every 2 weeks	Days of collecting images	Days of operating tractors
Delay of the report	The day following the day of collecting images	3 to 5 business days after the day of collecting images	After the necessary processing time	After the necessary processing time
Initial costs	0	0	One million yen	Three million yen?
Operational costs	Based on the area number of sheets is unlimited	Based on the area as well as the number of sheds	Commissioned personnel expenses based on the required numbers of days	Labor costs for farm workers based on the required numbers of days
Merits	Since it is delivered through the website on the day following the day of collecting images, high-frequency medium-resolution data can easily be confirmed	Wide area collection of images (77 km) allows comparison of wide ranges	Data can be obtained even when it is cloudy	The data can be available whenever you want to see it even at night or on a cloudy day
Demerits	Data cannot be acquired if there are clouds. Since the influence of clouds cannot be eliminated, users have to get used to identifying the necessary data	Data cannot be acquired if there are clouds. Even if it is sunny, it may not be possible to collect images, causing erroneous results. The information becomes old when it is delivered to users. An increase in the number of sheets increases the cost of images	It takes considerable efforts from data acquisition to analysis. Outsourcing of the processes increases the costs. Noise may occur as a result of mosaic processing	A vehicle has to be operated to obtain data, which takes time and effort. Morning dew· may cause noise

[a]Since the data on Crop spec is based on verbal information, it may be inaccurate

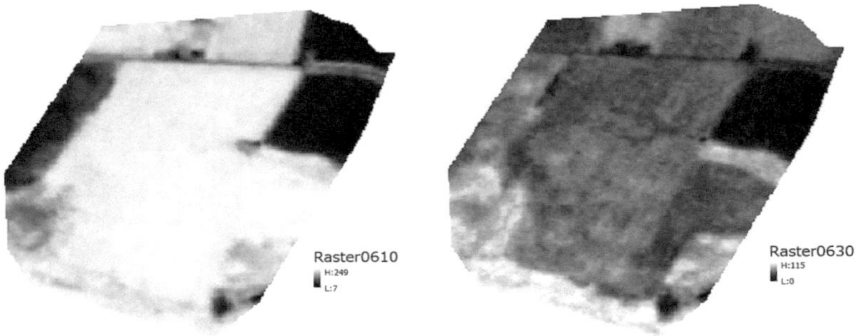

Fig. 4.9 Planet NDVI 20180610 (left) and Planet NDVI 20180630 (right)

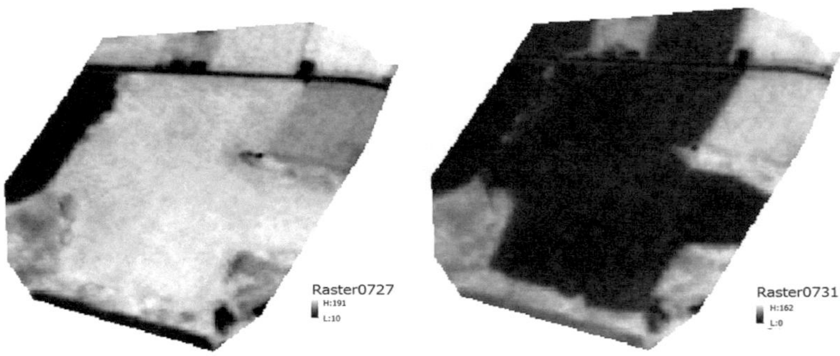

Fig. 4.10 Planet NDVI 20180727 (left) and Planet NDVI 20180731 (right)

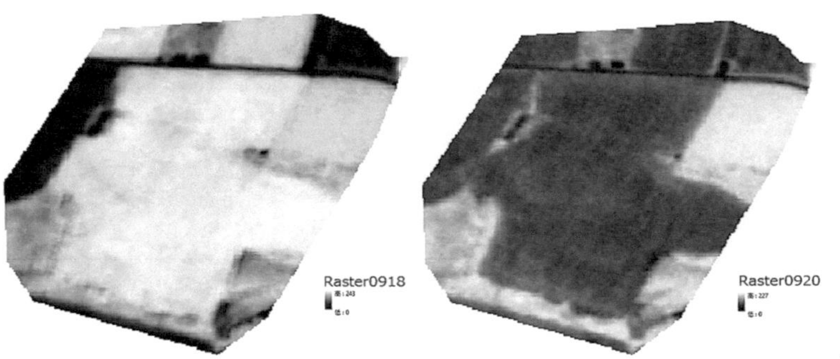

Fig. 4.11 Planet NDVI 20180918 (left) and Planet NDVI 20180920 (right)

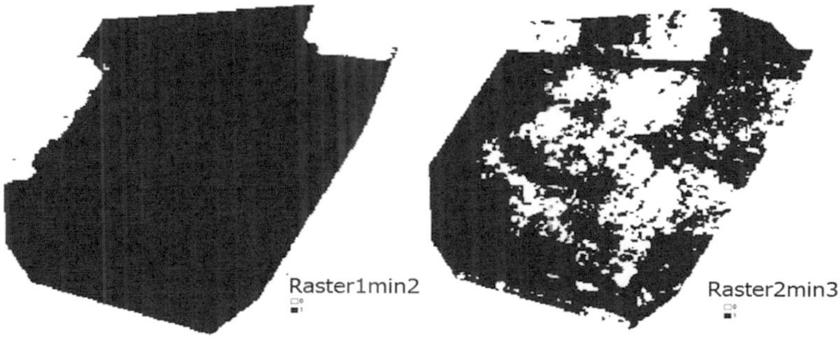

Fig. 4.12 Comparison of BNDVI images among first, second, and third harvesting period

4.6 Conclusion and Future Works

UAVs equipped with RGB and near-infrared cameras that take Blue Normalized Difference Vegetation Index (BNDVI) are used to detect overfertilized (less fertilized) area by analyzing two images crossing harvest period. A slope map generated from digital elevation model (DEM) has a relationship with less fertilized area. Fertilizer does not remain if the slope is steep. To prevent this, farmers change the intensity of fertilizing. To acquire DEM, UAVs have a great advantage. The author has tried to acquire height maps of grassland by using DEM. However, height data value is not stable and every time the height changes. To solve this problem, the author thought of using RTK system. The comparison between height of grassland and intensity of NDVI is expected. With a support of Space-Agri company, the author acquires remote sensing data for study area and highlights less fertilized area. Remote sensing data is influenced by light cloudiness. Acquisition of remote sending data is cost-effective compared with that of UAVs. Fixed wing UAVs can cover a wider range of area. The author has used Parrot Disco fixed wing UAVs. Parrot provides Parrot AG equipped with Sequoia camera. However, the cost of Parrot AG is almost three or four times that of Parrot Disco with RGB camera. To reduce the cost, the author is building a handmade UAV with support from a Malaysian who has experience from a German UAV building company and modifies a normal camera to NDVI camera by using IR filter (Drone Rice Project 2019). Using handmade fixed wing can reduce the acquisition cost. The combination of remote sensing data and UAVs can be a complimentary relationship. The author will start to use Sequoia camera from 2019 and use this camera for calibration. The advantage of this camera can capture red-edge bands. It gives the changes in the reflectance in between red band and near-infrared band. If we see the reflectance graph of vegetation, then we can notice that there is a sharp change in the reflectance value from red to infrared band; therefore the missing portion of the reflectance can be covered by red-edge band. This band can be useful to give information about the change in the phenological properties of vegetation as well as species level changes.

The author has got a fertilizer delivering UAV license, the DJI-manufactured MG-1 (MG-1 2019). According to a survey from the dairy farmers, grassland should be added with fertilizer of 100–130 kg per 10-m by 10-m area per year. Making use of maps produced through this experiment, variable fertilizing is useful. Especially step slope where a tractor cannot enter, using UAVs is very useful and saves lives of dairy farmers. The author obtains precise digital elevation model and then delivers fertilizer where a tractor can't enter.

Acknowledgment Owing to the great support of the Iwamoto Farm in the eastern part of Hokkaido, Japan, this research is a work in progress. This work is supported by JSPS KAKENHI Grant Number JP16K07979 and JP16K00658.

References

Drone Rice Project (2019) https://drone-rice.jp/

GIS Geography What is NDVI (Normalized Difference Vegetation Index)? https://gisgeography.com/ndvi-normalized-difference-vegetation-index/. Accessed on 18 Feb 2019

GreenSeeker handheld crop sensor https://agriculture.trimble.com/precision-ag/products/greenseeker/. Accessed 26 Feb 2019

Inoue H, Uchiyama S, Suzuki H (2014) Multicopter aerial photography for natural disaster research, NIED research report

IR-Pro (2017) https://www.irprostore.com/

Lu S, Inoue S, Shibaike H, Kawashima S, Yonemura S, Du M (2015) Detection potential of maize pollen release stage by using vegetation indices and red edge obtained from canopy reflectance in visible and NIR region. J Agric Meteorol 71(2):153–160

Matsumura and Inoue (2018) Unmanned Aerial Vehicle for fertilizer management and human health. Med Res Arch 6(4):1–7

MG-1 drone (2019) https://www.dji.com/jp/mg-1s. Accessed on 31 Mar 2019

Planet Company Ltd (2018) https://www.planet.com/. Accessed on 3 Jan 2019

Sequoia (2019) https://www.parrot.com/jp/ye-wu-yong-soriyusiyon/parrot-sequoia#parrot-sequoia -. Accessed on 10 Mar 2019

SpaceAgri Company Ltd (2018) https://www.space-agri.com/. Accessed on 3 Jan 2019

Chapter 5
Corn Height Estimation Using UAV for Yield Prediction and Crop Monitoring

Flavio Furukawa, Kenji Maruyama, Youlia Kamei Saito, and Masami Kaneko

Abstract Precision agriculture is improving agriculture management worldwide through technologies such as global navigation satellite system (GNSS), robotics, sensors, and variable rate applications. Geographic information system (GIS) and remote sensing are fundamental techniques for precision agriculture, providing different types of information: plantation layouts, crop health, and plant growth stages; these tools can provide information to farmers quickly. The popularization of unmanned aerial vehicles (UAV) made those aircraft more affordable and easy to use, providing information with high spatial and temporal resolutions. This study aimed to predict corn crop yield through corn height estimation generated by 3D photogrammetry based on structure from motion technology. The UAV data were taken in 14-flight campaign to acquire 3D; red, green, and blue (RGB); and normalized difference vegetation index (NDVI) data for 5 months in 2017 and compared with the ground data obtained in the harvest in middle October of the same year. The methodology allows understanding the whole field, while other methods are based on sample data, showing it to be more convenient since it is less time-consuming. Considering only the UAV height estimation (UHE) variable, the prediction reached an R-squared value of 0.51 with dry grain yield at the beginning of August and allowed plant height monitoring after NDVI saturation, presenting a high potential for yield prediction and crop monitoring.

Keywords UAV · Photogrammetry · Height estimation · Yield prediction · Remote sensing

---

F. Furukawa (✉) · K. Maruyama · Y. K. Saito · M. Kaneko
Department of Environmental Sciences, Rakuno Gakuen University, Ebetsu-shi, Hokkaido, Japan
e-mail: maruken@rakuno.ac.jp; kaneko@rakuno.ac.jp

R. Avtar, T. Watanabe (eds.), *Unmanned Aerial Vehicle: Applications in Agriculture and Environment*, https://doi.org/10.1007/978-3-030-27157-2_5

5.1 Introduction

By 2050, the world's population will reach 9.1 billion, an increase of 34% compared to 2009, mostly in developing countries, with 70% of the population living in urban areas (FAO 2009). To feed this entire population, an increase of 70% in food production is necessary. As an example, the annual cereal production has to reach 3 billion tons by 2050; in 2009 the annual cereal production was 2.1 billion tons. Despite that fact, another challenge is to feed this population sustainably and increase food production considering the safety and conservation of natural resources. The sustainability concept is based on the principle that the needs of the present are met without compromising the needs of the future (Spiertz 2009). Pretty (2008) introduced that sustainability demands a development in agricultural technologies and practices that do not affect the environment, leading to growth in food quality and production with minimum side effects against the environment, which is called precision agriculture.

The term precision agriculture can be explained as agriculture techniques that help the producers to take better decisions per unit area of land per unit time which increase their benefits (McBratney et al. 2005). It means an increase of quality and/or quantity of production, reducing environmental hazards derived from excessive input (fertilizers and pesticides) application, increasing their usage efficiency, and even reducing them, through variable management practices (Tang and Turner 1999). As Bongiovanni and Lowenberg-Deboer (2004) instanced, the concepts between sustainability and precision agriculture are linked, using agricultural machinery coupled with global navigation satellite system (GNSS), the application of fertilizers and pesticides where and when they are needed became possible. Called as variable rate application technique – a concept published by the University of Illinois back in 1929 (Sawyer 2013) – the usage of fertilizers and pesticides more accurately on the crops generates more possibilities to reduce their usage, reaching only the interested area, improving plant health, reducing costs, and collaborating to reduce environmental damages.

Geographic information system (GIS) along with the improvements in spatial and temporal resolutions of remote sensing technologies in different platforms (satellites and aircraft) strengthens the suitability for precision agriculture technique (Matese et al. 2015). These developments make it capable to acquire fairly reliable field data through a nondestructive method, delivering measurement data through the electromagnetic spectrum, allowing the assessment and monitoring of the crop chlorophyll status, spatial distribution of the crop, addressing important issues such as crop growth monitoring, vegetation stress detections, different predictions and improvement of crop management practices (Haboudane et al. 2008).

Unmanned aerial vehicle (UAV) is a remote sensing platform that has been highlighted due to the low cost and high spatial and temporal resolutions it offers (Salamí et al. 2014). UAVs, also known as drones, are being widely used for precision agriculture purposes, capturing imagery for plant/ crop analysis and acquiring information on soil water holding and irrigation systems (Ipate et al.

2015). This technology is based on an aircraft remotely controlled by a human or computer, coupled with a digital camera and/or different types of sensors (multispectral, thermal, Light Detection And Ranging – LiDAR) to acquire data of a specific scenario. Using photogrammetric techniques, those acquired data can result in image orthomosaics and point dense clouds (Rokhmana 2015).

Vegetation indexes are one of the products derived from UAVs coupled with a multispectral sensor and multispectral satellites. The normalized difference vegetation index (NDVI), developed by Rouse et al. (1973), is a widespread vegetation index in the precision agriculture and natural environment fields. This index is calculated from the normalized total reflectance from infrared and red (IR + red) bands, and it is used to estimate vegetation health condition and vegetation changes through the time. The study made by Huang et al. (2014) also found a strong correlation between NDVI data and crop yield, being possible to establish crop yield estimation models, even without any historical crop yield records. Bendig et al. (2015) mentioned that the usage of UAV to acquire different indexes and plant height is an exceptional way to acquire data for agriculture purposes, giving a high temporal and spatial resolution with high-cost benefit.

Another result obtainable through UAV platform is the 3D point cloud, an output that can be obtainable through laser systems (LiDAR) or 3D photogrammetry technology. Structure from motion technique is a low-cost 3D photogrammetry technology that recreates a structure by overlapping a set of offset images, using the principles of stereoscopic photogrammetry (Westoby et al. 2012). The stereoscopic technique creates the illusion of depth, simulating the human binocular vision (Ortis et al. 2013), which sees the same scene with a slightly different angle from each eye, generating a 3D point cloud. Each point has its own position information (X, Y, and Z) representing different structure dimensions in a 3D environment, enabling different analyses of the 3D model.

The present research aims to estimate the height of a corn crop field, using DJI Phantom 4 Pro, a nonmodified commercial UAV, through 3D photogrammetry technology for corn crop yield estimation and growth monitoring. Yin et al. (2011) affirm that early to mid-season crop plant height is reliable as a predictor of corn yield, which has a strong relationship with nitrogen application rate, confirming with Raun et al. (2001) that refers that grain yield goals is the most reliable method to estimate preplant fertilizer nitrogen rates. Along with that, the authors aim to understand the plant height growing rate comparing with the widely used vegetation index on the Precision Agriculture field, the NDVI.

5.2 Materials and Methods

5.2.1 Site Description and Management

The experiment was conducted on a field located southwest of Rakuno Gakuen University, in Ebetsu, Hokkaido, Japan, between the geographic coordinates

Fig. 5.1 Study area in the southwest side of Rakuno Gakuen University, Hokkaido, and the field grids with their respective sizes expressed in meters

43°4′4.81″N, 141°30′26.81″E, and 43° 4′4.03″N, 141°30′27.67″E. The field of 363.48 m^2 was divided into six rows and six columns, totalizing 36 grids, labeled from E1 to E36; the limits were defined with wood stacks and strings. In the first five rows, each grid measured 3.8 m × 2.6 m, and in the last row, each grid measured 4.3 m × 2.6 m (Fig. 5.1). The shapefile of the grid was transcripted through ArcMap with the red, green, and blue (RGB) georeferenced UAV data from May 26 as a reference.

The seeding of the hybrid corn 36B08 from Pioneer Hi-Bred International was made on May 1, 2017, using a manual seeder, each grid receiving a different treatment during the season. Harvest was made on October 14, 2017. For data comparison, ten samples of each grid were collected on harvest day, obtaining the average of plant height of each grid in centimeters, total dry matter in g/m^2, and dry grain yield in g/m^2.

5.2.2 Materials

The following equipment and computer resources were used to collect and process data:

- DJI Phantom 3 Professional, quadcopter UAV, coupled with a 3.97 mm NDVI lens from Peau Productions Inc. used to acquire NDVI data. This lens allows all light wavelengths to reach the sensor, separated into infrared and red bands on the blue and red channel, respectively.
- DJI Phantom 4 Pro, quadcopter UAV, used to acquire RGB and 3D data.

- Pix4D Capture application used to design a specific path for 3D models, exporting all images in JPEG format, and to control the flight for both UAVs remotely.
- Map Pilot application, from Maps Made Easy, used to shoot raw images in DNG format, which is essential to perform reflectance calibration on the acquired IR + red band data from NDVI lens, to design flight path and control the flight for both UAVs remotely.
- Mapir Camera Reflectance Calibration Ground Target used for the calibration of the IR + red images from DJI Phantom 3 Professional coupled with the NDVI lens.
- QGIS 2.14 with Mapir Processing Plugin was used to calibrate the image digital number values into reflectance values.
- Agisoft Photoscan Professional software used to generate orthomosaics and point dense cloud.
- CloudCompare V2.8.1 Stereo software used to edit the dense point cloud created with Agisoft Photoscan Professional software.
- ENVI 5.4 software used to manage multispectral data, perform remote sensing analysis, and compare datasets from different remote sensing platforms.
- ArcMap 10.5 software used to perform spatial data processing and compare datasets from different remote sensing platforms.

5.2.3 *Data Acquisition*

For data acquisition, the UAV DJI Phantom 3 Professional with NDVI lens and DJI Phantom 4 Pro were used. Forty-two flights were performed in 14 weeks from May to September of 2017 (Fig. 5.2); in each campaign, three flights were performed to acquire NDVI, RGB, and 3D data. An average of 317 3D images, 160 RGB images, and 214 IR + red images were collected per day.

The NDVI data was acquired using the Phantom 3 Professional with NDVI lens and Map Pilot application, flying at 50 m above the ground, angle of the camera at 90°, setting an overlap of 90% and side lap of 70%, with white balance set at auto mode, ISO 100, shutter speed of 1/1000, and DNG file format. After flying over the

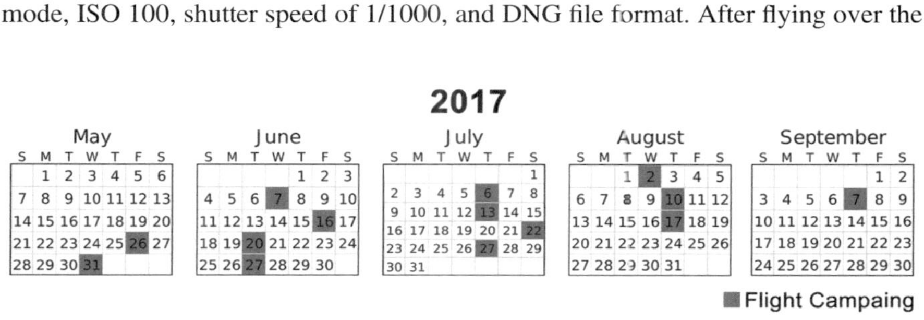

Fig. 5.2 Data acquirement dates, between May and September 2017

study field, a shot of Mapir Camera Reflectance Calibration Ground Target was taken to calibrate the IR + red photos.

The RGB data was taken using a DJI Phantom 4 Pro and Map Pilot application, flying at 50 m above the ground, using camera settings at auto mode, with an overlap of 90% and a side lap of 70%, angle of the camera at 90° and DNG file format.

For the 3D data, the same DJI Phantom 4 Pro was used with Pix4D Capture application, flying at 50 m above the ground, angle of the camera at 70°, setting the overlap at 90% and side lap at 70%, JPG file format, using the camera at automatic settings, with trigger mode set at fast mode.

5.2.4 Data Processing

The acquired data were processed through structure from motion technology, which consists of advanced algorithms to create 3D structures from overlapped 2D imagery. It is based on stereo vision principle, which compares the same scene from different angles to match common points.

Orthomosaics and point dense cloud were created using Agisoft Photoscan Professional and medium settings for all processes (to align photos, create dense cloud and mesh) and applying the WGS84/UTM zone 54 N projection for every output data. The orthomosaics generated had a resolution of approximately 1.5 cm, and data were georeferenced in ArcMap using a WorldView-2 (WV2) imagery from June 13, 2017, as a reference.

5.2.4.1 NDVI

After collecting the images obtained with the Phantom 3 Professional and the NDVI lens, the data were processed to reduce lens distortion and converted from DNG file to TIFF file through Mapir Processing Plugin for QGIS to be processed on the photogrammetry software for more accurate NDVI datasets. The orthomosaics created through Agisoft Photoscan Professional were calibrated to reflectance values using the Mapir Camera Reflectance Calibration Ground Target image on the same QGIS plugin.

Due to the different climatic conditions when the images were obtained, the NDVI data showed significantly different values for the same area in a short period of time. In order to compare these values, the data went through a simple normalization. The roof of a building was used as a reference for this process, and an algorithm created in Map Algebra tool on ArcMap was performed. After this normalization, it was possible to compare the NDVI behavior through the weeks for each grid.

To assess the accuracy of the NDVI lens, comparisons between the UAV data and radiometrically corrected satellite data were made considering the average of each grid in three periods. The high spatial and spectral resolution imageries (50 cm of resolution) from WorldView-2 (WV2) and GeoEye-1 (GE1) are operated by DigitalGlobe. The comparisons were between the UAV and WV2 data from July

13, the UAV data from August 17 and the WV2 data from August 23, and the last one using the UAV data from September 7 and GE1 data from September 1. The data were preprocessed in ENVI, converting digital number values in radiometric values. In the satellite data from August 23, a thin haze was detected affecting the comparison with UAV data from August 17.

Green coverage classification was also performed to extract NDVI values only from the vegetation to understand the influence of different variables such as soil and shade on the NDVI values. Using ArcGIS Interactive Supervised Classification tool, the data were classified in two classes: no vegetation and vegetation. A shapefile was created to extract NDVI average for the vegetation of each grid using the Zonal Statistics as Table tool.

5.2.4.2 Height Estimation

All point cloud datasets generated by Agisoft Photoscan Professional through 3D photogrammetry technique mentioned before were aligned with CloudCompare software, through Cloud Registration function with data from May 26 as a reference for alignment, setting the final overlap in 30% and the random sampling point to 100,000. In ArcMap, one last dataset was created for each dense point cloud, exported into a raster file for comparison, and georeferenced using WV2 imagery from July 13 to spatially align all rasters. The Spatial Analyst toolbox was required to calculate the UAV height estimation (UHE) from each week using the Map Algebra tool, applying a simple subtraction with data from May 26 as the ground reference, since there were no plants at the time.

5.3 Results

5.3.1 NDVI Assessment

The comparison between NDVI lens with high spatial and spectral resolution WV2 and GE1 imagery resulted in a significant correlation with both satellites' imagery. Considering the average of each grid, on July 13, the NDVI data from the UAV and the WV2 showed an R-squared value of 0.76. Comparing the data obtained with UAV on August 17 and the data from the WV2 imagery taken on August 23, an R-squared value of 0.527 was found, and this relatively low value on the WV2 imagery was due to the thin haze mentioned before. The last data acquired through UAV were on September 7, and compared to GE1 imagery from September 1, an R-squared value of 0.659 was presented.

The green coverage classification assessment, which was made to detect the interference of the soil on NDVI values, showed the same pattern of NDVI average without extracting the vegetation through classification (Fig. 5.3a), having a correlation of 0.98 between them (Fig. 5.3b).

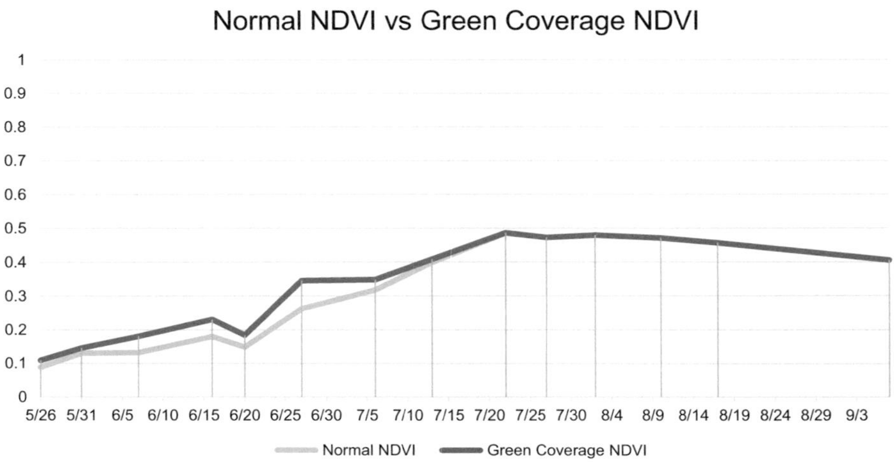

Fig. 5.3a Normal NDVI and green coverage through time

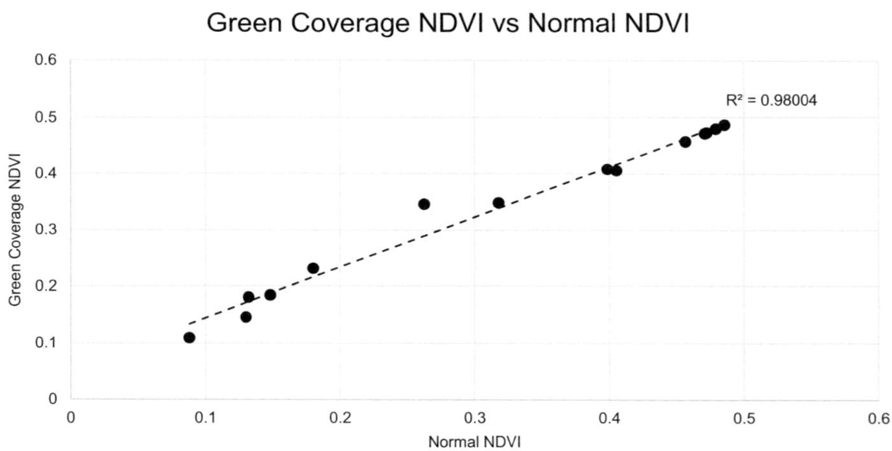

Fig. 5.3b Correlation between green coverage NDVI and normal NDVI

5.3.2 *UAV Height Estimation Assessment*

Considering the saturation on UAV height estimation average from August 17 to September 7, the correlation of each grid average obtained between the UAV height estimation and the ground measured height in October 14 was 0.87 (R-squared) (Fig. 5.4).

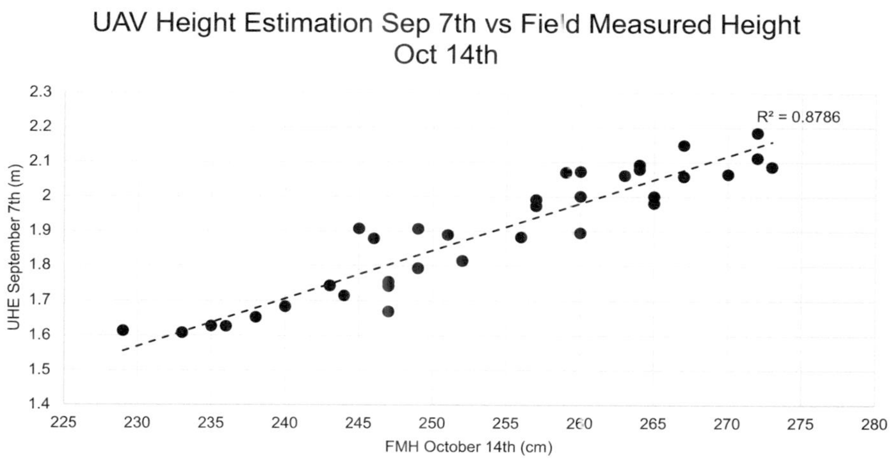

Fig. 5.4 Correlation between UAV height estimation from September 7 and field measured height (FMH) from October 14

5.3.3 NDVI and UAV Height Estimation Through Time

NDVI and UAV height estimation averages presented the following characteristics through time (Fig. 5.5):

From May 26 to June 20, the UAV height estimation average of each grid presented negative values, reaching the lowest value at June 16 – around minus 7 cm. On June 27, the UAV height estimation average had the first positive value. Contrarily, NDVI average showed a constant increase from May 26 to July 22; only on June 20 a lower value was found contradicting the increase pattern.

From June 27 to August 17, UAV height estimation average showed a consistent increase, while the NDVI average reaches its saturation on July 22.

UAV height estimation average showed a saturation after August 17, maintaining the same value until September 7 – 3 weeks later. NDVI average presented a small decrease from August 2, reaching the 0.406 value on September 7.

5.3.4 Field Data

The data measured on the field (Table 5.1) showed a correlation of 0.43 between dry matter and field measured height (Fig. 5.6) and a correlation of 0.47 between dry grain and field measured height (Fig. 5.7), considering each grid average.

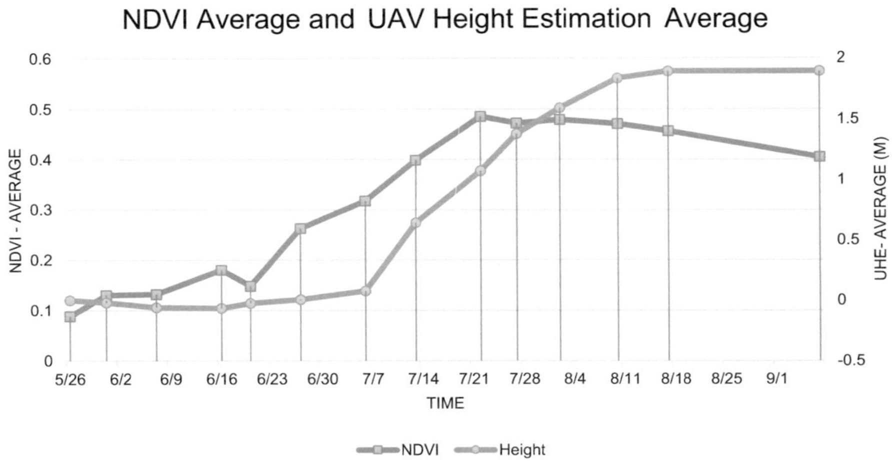

Fig. 5.5 NDVI average and UAV height estimation average through time

5.3.5 Correlations Through Time

The correlation between UAV height estimation through time and dry matter showed its highest R-squared value of 0.479 on August 2 (Fig. 5.8). Comparing UAV height estimation with dry grain yield through time, the highest correlation found on this same day was an R-squared value of 0.510 (Fig. 5.9).

The same happened to NDVI when compared with dry matter (Fig. 5.10) and dry grain yield (Fig. 5.11), and all the correlation values were under 0.5.

The correlation between UAV height estimation through time and the field measured height, obtained on October 14, presented a significant correlation from August 2, with an R-squared value of 0.68 (Fig. 5.12). From the middle of August, this correlation became more significant and the model could explain almost 90% of these variables.

5.3.6 Crop Monitoring

Over 14 weeks, the UAV height estimation showed the behavior expressed in Fig. 5.13, with a resolution of 2 cm per pixel with higher values displayed in magenta and lower values in green.

NDVI obtained through UAV presented the behavior shown in Fig. 5.14, with 2 cm per pixel, with higher values expressed in green and lower values in yellow.

Table 5.1 Data measured on the field on October 14

Code	Ground measured height (GMH) (cm)	Dry matter (g/m^2)	Dry grain yield (g/m^2)
E1	247.00	1202.48	527.34
E2	257.00	1270.21	630.46
E3	270.00	1322.64	656.09
E4	265.00	1456.59	680.74
E5	251.00	1418.44	658.36
E6	252.00	1445.53	680.14
E7	238.00	1115.35	468.59
E8	264.00	1484.55	727.44
E9	272.00	1527.42	748.28
E10	272.00	1242.88	599.99
E11	273.00	1477.35	743.18
E12	260.00	1586.01	778.88
E13	236.00	979.08	410.28
E14	263.00	1227.48	595.50
E15	267.00	1364.64	706.65
E16	267.00	1369.74	630.04
E17	260.00	1626.91	796.13
E18	247.00	1393.03	659.01
E19	244.00	1135.32	542.26
E20	259.00	1095.74	472.20
E21	260.00	1388.70	702.17
E22	246.00	1135.26	510.25
E23	249.00	1281.22	630.51
E24	235.00	1182.28	534.21
E25	247.00	1039.27	417.94
E26	264.00	1211.53	581.44
E27	257.00	1135.21	471.92
E28	240.00	1005.37	409.14
E29	256.00	1375.95	697.22
E30	229.00	978.27	399.63
E31	247.00	1056.25	509.55
E32	265.00	1451.59	667.39
E33	245.00	1147.63	552.02
E34	243.00	1117.47	495.68
E35	249.00	1456.77	747.82
E36	233.00	1122.25	470.97

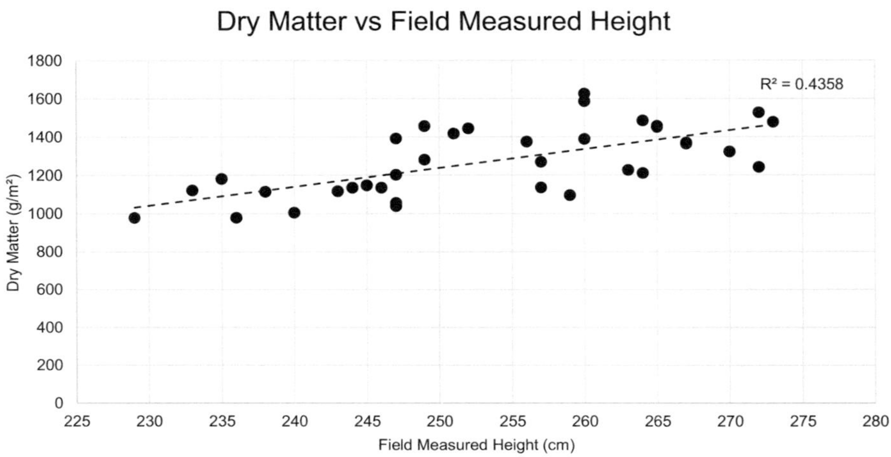

Fig. 5.6 Correlation between dry matter and field measured height

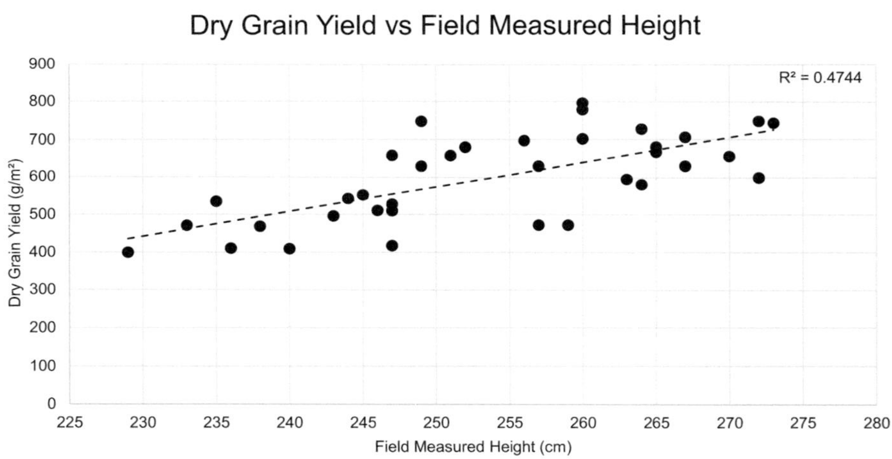

Fig. 5.7 Correlation between dry grain yield and field measured height

5.4 Discussion

The high correlation between satellite data (WV2 and GE1) and UAV data taken with NDVI filter presented the capability of the NDVI lens to estimate NDVI using a low-cost system with high spatial resolution (around 2 cm), giving enough details for vegetation analysis.

Since the NDVI average values of each grid were used in this study, a high correlation between NDVI average value obtained with green coverage classification and the NDVI average without the green coverage classification was presented, with

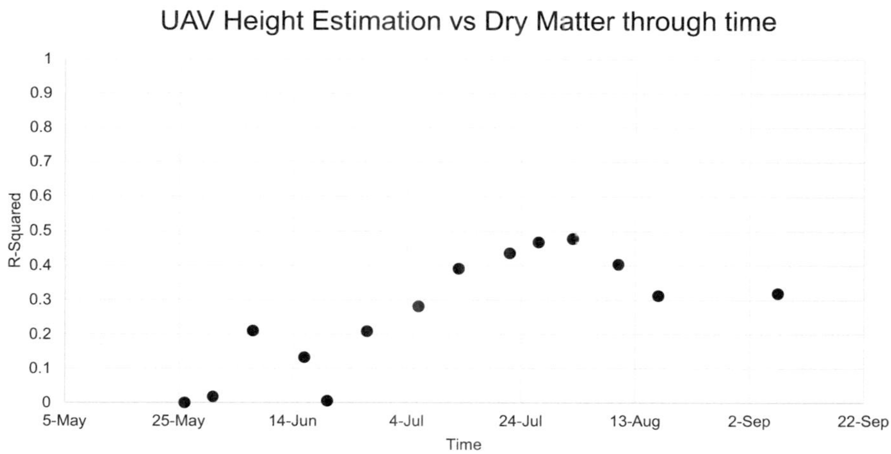

Fig. 5.8 Correlation between UAV height estimation and dry matter through time

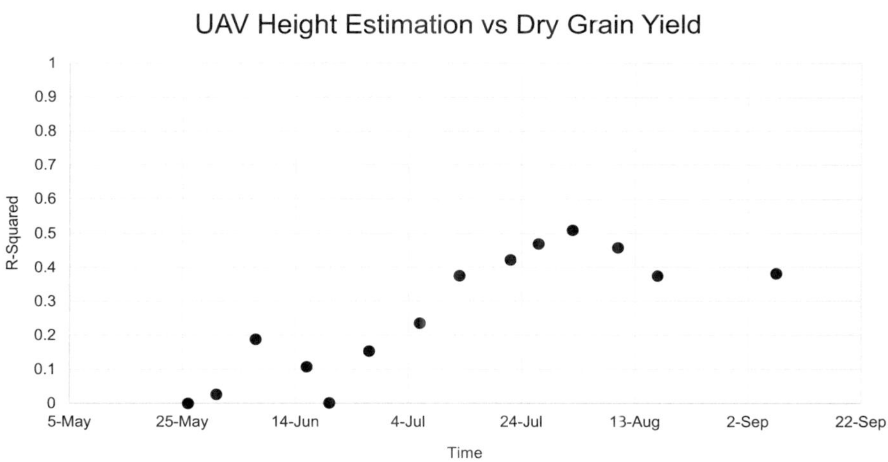

Fig. 5.9 Correlation between UAV height estimation and dry grain yield through time

an R-squared value of 0.98. The NDVI without green coverage classification values were used in this study because of its simplicity, requiring no imagery classification.

NDVI average obtained through UAV displayed a continuous growth until July 22 (when reached its saturation), with an exception on June 20, when the NDVI average value was lower than the previous date, June 16. This fact can be explained because of the soil influence on the NDVI average value (Huete 1988), and the same happened with NDVI obtained through green coverage classification, since the plant

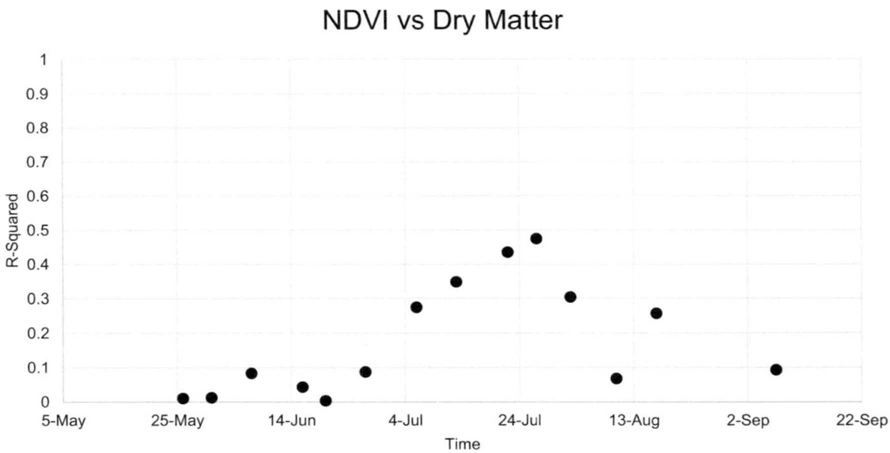

Fig. 5.10 Correlation between NDVI and dry matter through time

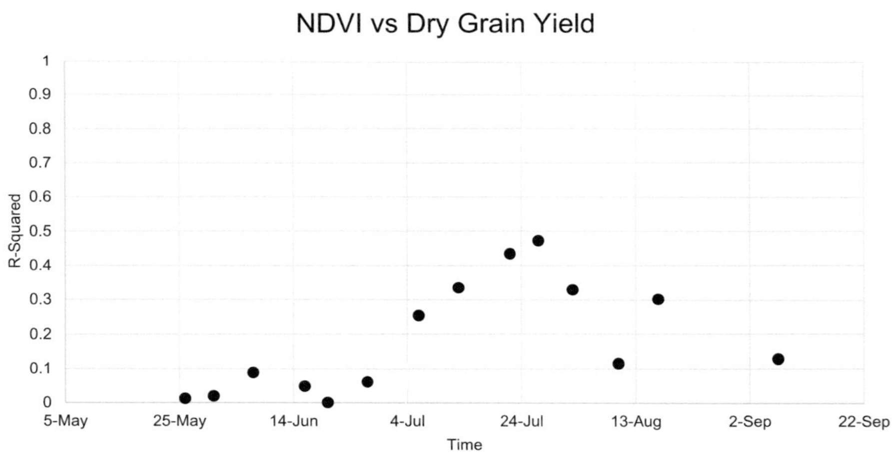

Fig. 5.11 Correlation between NDVI and dry grain yield through time

is very small, the classification algorithm allowed some soil inside the vegetation class. On June 16 and June 27, the soil moisture was higher than on June 20, which changed the color of the soil (Fig. 5.15) and occurring lower values.

Apart from that, NDVI average data obtained through UAV showed to be very sensitive on the early stages of the crop, enabling the ability to follow the growth of the crop. As Strachan et al. (2002) quoted, the plant growth is a function of nitrogen and water availability, and understanding the crop growth on early stages provides more chances to the producer to decide which management is adequate for the situation, improving crop health and enabling the capacity of yield growth monitoring.

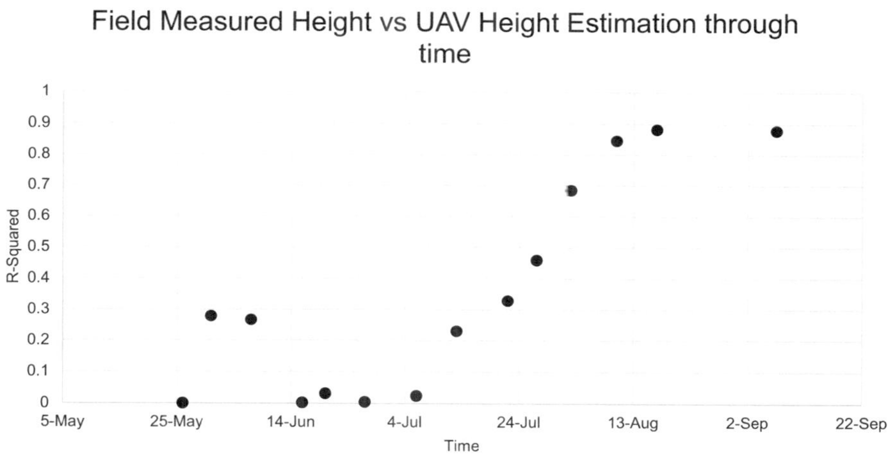

Fig. 5.12 Correlation between field measured height and UAV height estimation through time

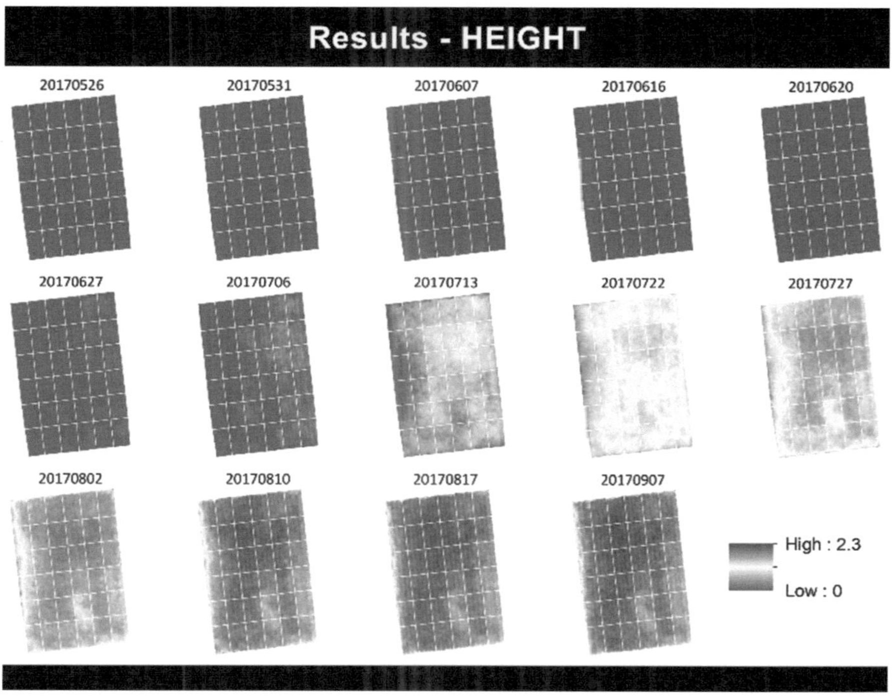

Fig. 5.13 UAV height estimation throughout time

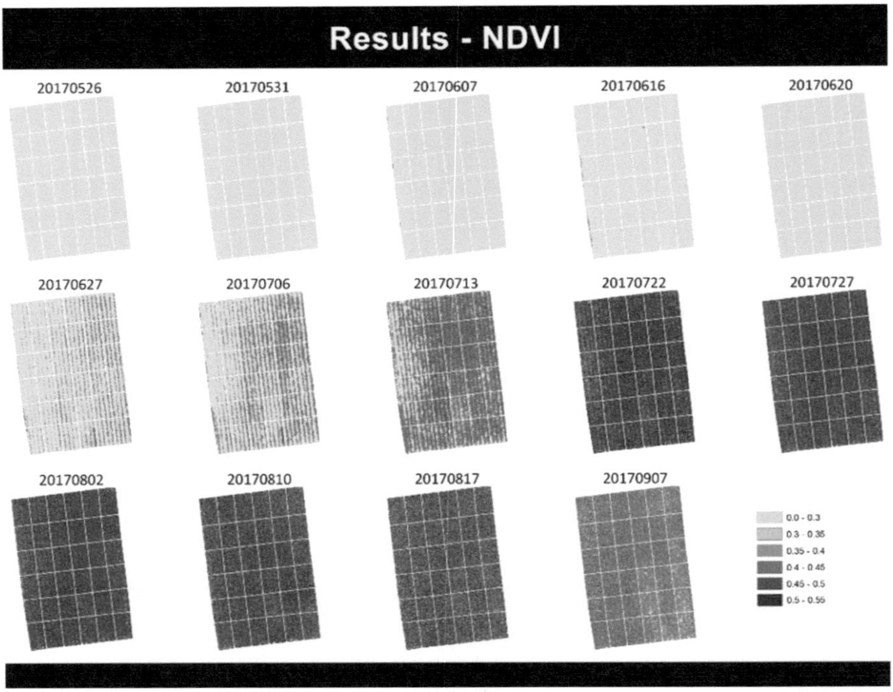

Fig. 5.14 NDVI from UAV throughout time

Fig. 5.15 Different soil moistures on June 16, 20, and 27

Conversely, UAV height estimation average presented negative values on early stages, until June 20. This may occur for two reasons: the plants' small size which cannot generate an accurate dense point cloud and the low accuracy of the cloud alignment through CloudCompare software. 3D photogrammetry struggles to determine thin or small objects, and according to a study by Javernick et al. (2014), the surface vertical error was 10 cm, using the same photogrammetry software (Agisoft Photoscan Professional), and also the camera light exposure influences on the final 3D point cloud (Blizard 2014). Therefore, the methodology and/or parameters behind point cloud alignment were unsuitable for plant growth monitoring at early stages of the crop, and some adjustments to the methodology are needed, such as point cloud classification and noise reduction.

UAV height estimation average had an increasing development from June 27 to August 17, where it reached its saturation. Compared to NDVI average, the UAV height estimation average reached its saturation 4 weeks later, allowing plant growth monitoring after the NDVI average values' saturation and consequently a decrease in its value. Those data showed that even though the plant reached the maturity and/or saturation on NDVI, the plants keep growing in height, and even so, UAV height estimation enables late stage growth monitoring, when the plants are vulnerable to drought stress, nutrient deficiency, or any kind of damage such as hail.

Considering the constant value of UAV height estimation from August 17 to September 7, the significant correlation obtained with October 14 field measured height values indicates that the height estimation generated through 3D photogrammetry imagery from UAV can track plant height growth rate over time through estimation, since 3D photogrammetry technology presented some limitations acquiring real size information.

Nevertheless, field measured height and UAV height estimation start having a significant correlation from August 2, enabling crop height prediction 10 weeks before harvest. But in this specific case, the hybrid 36B08 from Pioneer Hi-Bred International had a small correlation between field measured height and dry matter and field measured height and dry grain yield, with correlation values lower than 0.5 (0.4358 and 0.4744, respectively), showing that for this specific hybrid, height has no strong relationship with corn yield. Along with that, the correlations between UAV height estimation and dry matter and UAV height estimation and dry grain yield also had values lower than 0.5, and the same happened with NDVI values, consistent with what was achieved with UAV height estimation but confronting with Huang et al. (2014) where a strong correlation between NDVI and crop yield using time series data from MODIS-NDVI was found.

5.5 Conclusions

To achieve the food demand by 2050, those agricultural technologies are presenting to be necessary, not only in food production but also to preserve the environment, enhancing the movement of sustainable agriculture. Precision agriculture techniques are showing their potential to improve agricultural practices focusing on environmental health and economic profitability, along with GIS and remote sensing technology for crop management and monitoring.

The main goal of this research was to find the correlation between height estimation using a low-cost remote sensing platform with a UAV and corn yield through time, to give enough information in the right time to the producer to take adequate decisions in crop management of all stages, including after harvest. The methodology applied allows understanding the whole field showing it to be more convenient since it is also less time-consuming, while other methods are based on sample data, showing it to be more convenient since it is also less time-consuming. Considering only the UAV height estimation variable, the prediction reached an R-squared value of 0.51 with dry grain yield at the beginning of August and allowed plant height monitoring after NDVI saturation, presenting a high potential for yield prediction and crop monitoring.

Further studies will be necessary to conclude that corn height estimation using a non-modified commercial UAV is suitable for corn yield estimation. Improvements on the methodology (data acquirement and processing) could be done; manual camera setting and white calibration are recommended to acquire 3D data since light exposure seems to affect the dense point cloud creation. Along with that, different corn varieties should be tested in order to understand the relationship between corn height and yield.

References

Bendig J, Yu K, Aasen H, Bolten A, Bennertz S, Broscheit J, Gnyp ML, Bareth G (2015) Combining UAV-based plant height from crop surface models, visible, and near-infrared vegetation indices for biomass monitoring in barley. Int J Appl Earth Obs Geoinf 39:79–87. https://doi.org/10.1016/j.jag.2015.02.012

Blizard B (2014) The art of photogrammetry: how to take your photos – tested. Adam savage's test. http://www.tested.com/art/makers/460142-art-photogrammetry-how-take-your-photos/. Accessed 06 Jan 2018

Bongiovanni R, Lowenberg-Deboer J (2004) Precision agriculture and sustainability. Precis Agric 5:359–387. https://doi.org/10.1023/B:PRAG.0000040806.39604.aa

FAO-Food and Agriculture Organization (2009) How to feed the world in 2050. Insights Exp Meet FAO 2050:1–35. https://doi.org/10.1111/j.1728-4457.2009.00312.x

Haboudane D, Tremblay N, Miller JR, Vigneault P (2008) Remote estimation of crop chlorophyll content using spectral indices derived from hyperspectral data. IEEE Trans Geosci Remote Sens 46:423–436. https://doi.org/10.1109/TGRS.2007.904836

Huang J, Wang H, Dai Q, Han D (2014) Analysis of NDVI data for crop identification and yield estimation. IEEE J Sel Top Appl Earth Obs Remote Sens 7:4374–4384. https://doi.org/10.1109/JSTARS.2014.2334332

Huete AR (1988) A soil-adjusted vegetation index (SAVI). Remote Sens Environ 25:295–309. https://doi.org/10.1016/0034-4257(88)90106-X

Ipate G, Voicu G, Dinu I (2015) Research on the use of drones in precision agriculture. UPB Sci Bull 77:263–274. https://www.scientificbulletin.upb.ro/rev_docs_arhiva/full4af_166038.pdf. Accessed 21 Sept 2017

Javernick L, Brasington J, Caruso B (2014) Modeling the topography of shallow braided rivers using structure-from-motion photogrammetry. Geomorphology 213:166–182. https://doi.org/10.1016/j.geomorph.2014.01.006

Matese A, Toscano P, Di Gennaro SF, Genesio L, Vaccari FP, Primicerio J, Belli C, Zaldei A, Bianconi R, Gioli B (2015) Intercomparison of UAV, aircraft and satellite remote sensing

platforms for precision viticulture. Remote Sens 7:2971–2990. https://doi.org/10.3390/rs70302971
McBratney A, Whelan B, Ancev T, Bouma J (2005) Future directions of precision agriculture. Precis Agric 6:7–23. https://doi.org/10.1007/s11119-005-0681-8
Ortis A, Rundo F, Di Giore G, Battiato S (2013) Adaptive compression of stereoscopic images. Springer, Berlin/Heidelberg, pp 391–399. https://doi.org/10.1007/978-3-642-41181-6_40
Pretty J (2008) Agricultural sustainability: concepts, principles and evidence. Philos Trans R Soc Lond Ser B Biol Sci 363:447–465. https://doi.org/10.1098/rstb.2007.2163
Raun WR, Solie JB, Johnson GV, Stone ML, Lukina EV, Thomason WE, Schepers JS (2001) In-season prediction of potential grain yield in winter wheat using canopy reflectance. Agron J 93:131–138. https://doi.org/10.2134/agronj2001.931131x
Rokhmana CA (2015) The potential of UAV-based remote sensing for supporting precision agriculture in Indonesia. Procedia Environ Sci 24:245–253. https://doi.org/10.1016/j.proenv.2015.03.032
Rouse JW, Hass RH, Schell JA, Deering DW (1973) Monitoring vegetation systems in the great plains with ERTS. In: Freden SC, Mercanti EP, Becker MA (eds) Third earth resources technology satellite-1 symposium. NASA, Washington, pp 309–317. https://ntrs.nasa.gov/archive/nasa/casi.ntrs.nasa.gov/19740022592.pdf. Accessed 25 Sept 2017
Salamí E, Barrado C, Pastor E (2014) UAV flight experiments applied to the remote sensing of vegetated areas. Remote Sens 6:11051–11081. https://doi.org/10.3390/rs61111051
Sawyer JE (2013) Contemporary issues concepts of variable rate technology with considerations for fertilizer application. J Prod Agric 7:195. https://doi.org/10.2134/jpa1994.0195
Spiertz JHJ (2009) Nitrogen, sustainable agriculture and food security: a review. In: Lichtfouse E, Navarrete M, Debaeke P, Véronique S, Alberola C (eds) Sustainable agriculture. Springer, Dordrecht, pp 635–651. https://doi.org/10.1007/978-90-481-2666-8_39
Strachan IB, Pattey E, Boisvert JB (2002) Impact of nitrogen and environmental conditions on corn as detected by hyperspectral reflectance. Remote Sens Environ 80:213–224. https://doi.org/10.1016/S0034-4257(01)00299-1
Tang C, Turner NC (1999) The influence of alkalinity and water stress on the stomatal conductance, photosynthetic rate and growth of Lupinus angustifolius L. and Lupinus pilosus Murr. Aust J Exp Agric 39:457–464. https://doi.org/10.1071/EA98132
Westoby MJ, Brasington J, Glasser NF, Hambrey MJ, Reynolds JM (2012) 'Structure-from-motion' photogrammetry: a low-cost, effective tool for geoscience applications. Geomorphology 179:300–314. https://doi.org/10.1016/J.GEOMORPH.2012.08.021
Yin X, Jaja N, McClure MA, Hayes RM (2011) Comparison of models in assessing relationship of corn yield with plant height measured during early- to mid-season. J Agric Sci. https://doi.org/10.5539/jas.v3n3p14

Chapter 6
Supporting Oil Palm Replanting Programs Using UAV and GIS in Malaysia

Pegah Hashemvand Khiabani and Wataru Takeuchi

Abstract As oil palm is cultivated in large-scale plantations, prior and post-planting operations are laborious and time expensive, and also manual implementation of some of these operations often results in incorrect measurements and information. Successful oil palm management in prior and post-planting operations requires effective techniques to collect precise information. Unmanned aerial vehicle (UAV) imagery is a low-cost alternative to field-based assessment but requires the development of methods to easily and accurately extract the required information. The individual oil palm tree detection and height assessment are important and labor-intensive tasks for large-scale plantations where trees taller than 15 m are being replanted due to the high cost of harvesting and low yield. Therefore, in this chapter we demonstrate a general work flow for individual oil palm detection and height assessment using UAV imagery. We explain local maximum (LM) and template matching (TM) techniques as two commonly used approaches in individual tree detection. The accuracy of each method was evaluated on 20 randomly selected plots on UAV image with recall, precision, and F-score method. The F-score for LM method is higher than TM methods which, respectively, are 0.83 and 0.60. From 17,252 oil palm trees in the 20 plots, LM algorithm could detect 1395 (almost 80%) and TM algorithm, 967 (almost 55%). Three hundred fifty-seven and 785 trees have been missed, respectively, in LM and TM approaches. In both cases, background vegetation incorrectly has been labeled as oil palm tree, where 141 and 322 objects have been falsely detected as oil palm tree in LM and TM approaches. LM worked better in almost all of the plots; however the performance decreased in highly dense area. Contradictory to TM approach, in LM approach, shadows did not affect the performance as it was reflected in precision value. In densely cultivated plots due to leaves overlapping of neighboring trees, algorithm failed, and also in sparsely cultivated plots, shadows caused some commission errors. Inherited distortion of

P. Hashemvand Khiabani (✉) · W. Takeuchi
Institute of Industrial Science, The University of Tokyo, Tokyo, Japan

R. Avtar, T. Watanabe (eds.), *Unmanned Aerial Vehicle: Applications in Agriculture and Environment*, https://doi.org/10.1007/978-3-030-27157-2_6

UAV image also caused some omission errors in TM approach. Individual detected tree with LM algorithm was used in the next part for oil palm height estimation overlaying with canopy height model.

Keywords RBG sensor · Structure from motion (SfM) · Canopy height model (CHM)

6.1 Necessity of Effective Oil Palm Management

As of 2018, the global palm oil consumption reached 30% of the global vegetable oil, and palm oil has turned into the most consumed vegetable oil (USDA 2019). *Elaeis guineensis* (known as oil palm) can produce about 10 tonnes of crude palm oil in unit of a hectare per year in suitable area (Corley and Tinker 2003). Due to high suitability of tropical regions for oil palm, this crop is extensively cultivated in Indonesia and Malaysia (Khai et al. 2017). This industry has been trying to satisfy the increasing global vegetable oil demand mainly by expanding area, which led to environmental concern about deforestation and biodiversity loss in tropical forest (Gaveau et al. 2016). International pressure forced this industry to shift to sustainable production of oil palm. Therefore, to maintain in international market, both countries have been committed to follow sustainable production of palm oil. The Roundtable on Sustainable Palm Oil (RSPO) is a nonprofit organization that focuses on sustainability within this sector through a certification program (Roundtable on Sustainable Palm Oil 2013). RSPO certification concerns about economically viable, environmentally appropriate, and socially beneficial operations in this sector. This certification considers a set of principles and criteria to evaluate the growers and millers. To improve oil palm yield per unit of hectare, appropriate implementation of best agricultural practices has been set as one of the RSPO's eight principles, in which both prior and post-planting operations are being considered. However, as oil palm is cultivated in large-scale plantations (Woittiez et al. 2017), these operations are laborious and time expensive. Manual implementation of some of these operations often results in incorrect measurements and information such as incorrect terrain operation or remote and inaccessible road locations.

Mechanization of labor-intensive field operations by new technologies is an urgent need for oil palm industry. New technologies such as global navigation satellite system (GNSS), geographic information systems (GIS), and remote sensing (RS) can help to improve implementation of both prior and post-planting operations. These technologies can be used in many aspects of prior planting operations such as land preparation, road and drainage construction, or post-planting operations such as monitoring of detailed individual tree attributes such as height and crown measurements and status of disease development.

RS is a powerful tool which provides repetitively accurate information at a large spatial coverage and has been successfully used in a significant number of studies to assist more effective oil palm management (Santoso et al. 2011; Tan et al. 2013; Chemura et al. 2015). However, the coarse resolutions of spaceborne data often cannot meet the requirements of detailed measurements, whereas the fine resolutions of spaceborne data are not publicly free available, and still the spatial resolution may not be high enough for precision agricultural managements (Tian et al. 2017). Also, in the case of oil palm plantations in Malaysia and Indonesia particularly, due to consistent presence of cloud cover, acquisition of cloud-free data on a specific time is a serious barrier (Gutiérrez-Vélez 2013).

The latest developments in unmanned aerial vehicles (UAVs) introduce these platforms as powerful alternatives to spaceborne remote sensing (Colomina and Molina 2014). Depending on the onboard sensor, UAVs provide various RBG, multispectral, hyperspectral, LiDAR, and thermal data, etc. at very high spatial resolutions and at flexible acquisition periods (Maes and Steppe 2019).

Precise location information is a critical requirement for all field operations where GNSS receiver on board UAV can provide high-precision location information in many of oil palm field operations such as locating new plantation and designing blueprints for replanting programs, terrain correction plans, etc. GNSS with increased number of satellite availability can be reliable sources where GPS cannot provide good accuracy (Dow et al. 2009).

Among post-planting operations, oil palm height measurement is one of the important operations, as usually trees taller than 15 m are being replanted due to the high cost of harvesting and low yield (Corley and Tinker 2003). However, this operation is labor and time expensive. UAV imagery has been used for tree high measurements in several studies (Mohan et al. 2017; Park et al. 2016; Wallace et al. 2014).

One of the commonly used approaches to detect height of individual tree from UAV data is to use canopy height models (CHMs). Recent progress on computer processing power and techniques makes it possible to build 3D models by matching same objects in multiple images or so-called structure from motion (SfM) technique (Westoby et al. 2012). Using ground control points (GCPs) and geo-referencing, the SfM technique provides digital elevation model (DEM) and digital surface model (DSM) which are the main components of CHMs. DEM represents the elevation of ground surface without considering height of surface objects; however, DSM represents the elevation of earth's surface. CHM is then the difference between DSM and DEM which represents the height of vegetations in agriculture applications.

Therefore, in this chapter, we introduce a general work flow for individual oil palm tree detection and height assessment using UAV imagery. Figure 6.1 shows the general work flow which is discussed in detail in different parts of this chapter. Parts A and B show two commonly used approaches (local maxima and template matching) in individual tree detection, and part C shows general work flow of accuracy assessment.

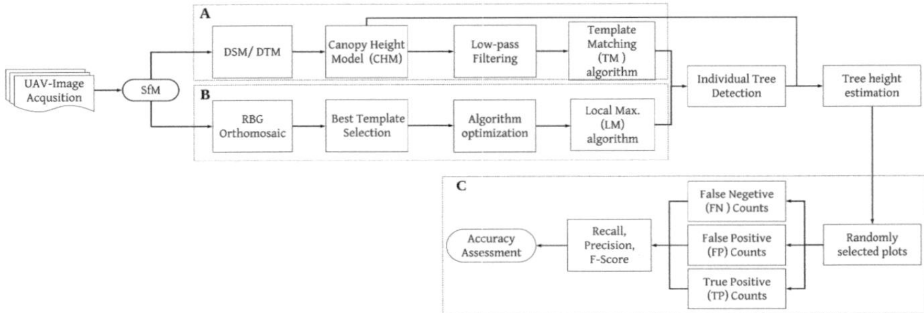

Fig. 6.1 General work flow for individual tree detection and height assessment using (**a**) local maximum approach and (**b**) template matching approach. Section C shows a general work flow of accuracy assessment

6.2 Oil Palm Individual Tree Detection Using UAV

This chapter uses UAV data captured with Canon PowerShot S100 commercial RGB digital camera on board UAV with focal length of 5.2 mm, 4000 × 3000 resolution, and pixel size of 1.86 × 1.86 um. The RGB sensor covered 6.97 km^2 of the area with 731 images and flight altitude of 292 m. The point clouds and CHM have been generated using the Agisoft Photoscan software where considering coordination of images, it provides point clouds and 3D construction of the area ("AgiSoft, L.L.C. PhotoScan Professional Edition"). The orthomosaic image was generated using GPS coordination and GCPs with the error of 1.54 pix. Figure 6.2 shows the orthomosaic RGB, DSM, DEM, and CHM images derived from UAV.

Two commonly used approaches for individual tree detection and height measurements are local maxima (LM) and template matching (TM). LM estimates individual tree height from CHM, where the algorithm finds pixel with maximum value among group of pixels on grayscale image (Pouliot et al. 2002). LM algorithm assumes that each tree has a single brightest pixel, and the rest of the tree crown area appear darker due to shading affect (Wulder et al. 2000). TM is another approach for individual tree detection which is used for locating of an object, where a template is a sub-image of a larger image and the algorithm tries to find occurrence of this template in the large image based on the template's boundary (Ahuja and Tuli 2013). Once individual trees have been detected, CHM is used to estimate the height of trees.

In the following we demonstrate LM and TM techniques for tree detection and height assessment.

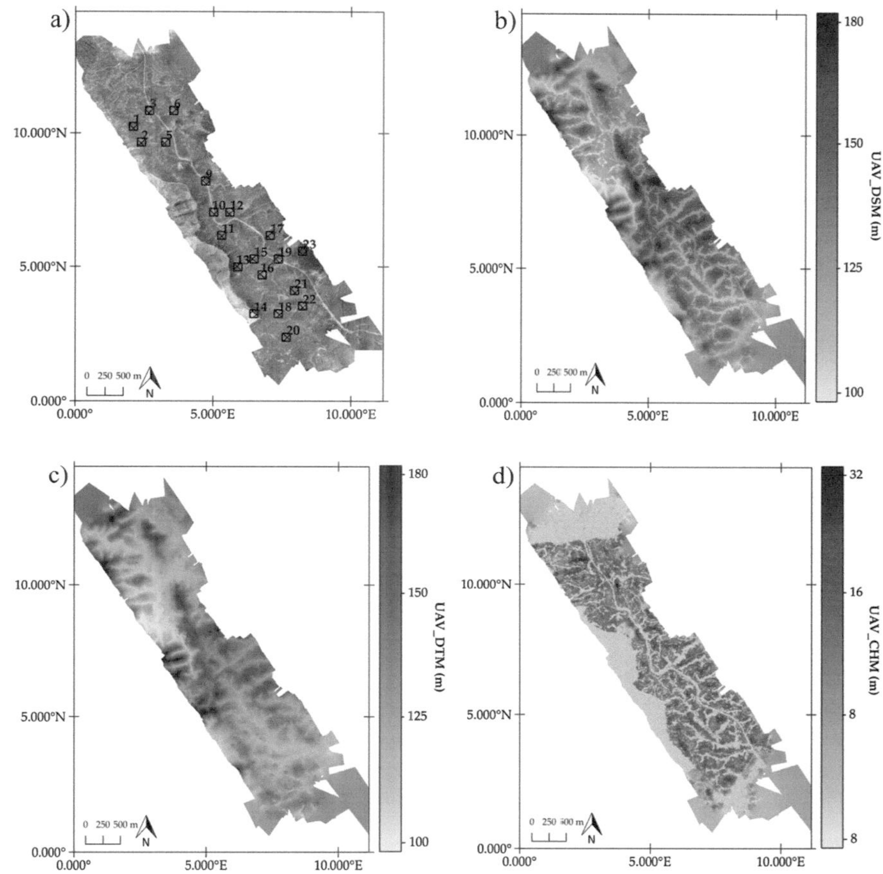

Fig. 6.2 UAV derived information layers. (**a**) RGB orthomosaic image. (**b**) UAV derived DSM. (**c**) UAV derived DTM. (**d**) UAV derived CHM

6.2.1 Application of Local Maximum Algorithm for Individual Tree Detection

In grayscale CHM image, the brightest pixels represent top of the tree crown. However, as sun and sensor angle effect pixel brightness, this brightest pixel may not always locate at the top of the crown. In LM approach, finding only one local maximum in each crown is an important and challenging task as the grayscale CHM images contain higher and lower frequency information where the higher frequency information shows the detailed crown structure information and noises and lower frequency information contains tree crowns and top canopy structure information. To reduce noises and improve LM algorithm performance, it is necessary to apply some low-pass filters in which high frequency information is blocked such as mean,

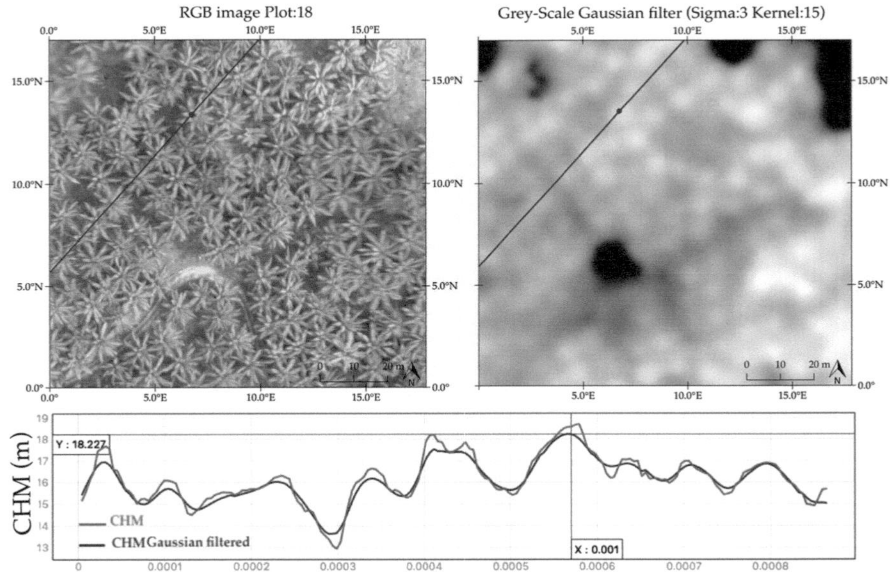

Fig. 6.3 Application of Gaussian filter for smoothing CHM

median, and Gaussian filters (Novotný et al. 2011). In this chapter the Gaussian filter is applied on the CHM image to filter out noises and unnecessary information. The Gaussian filter uses bell-shaped Gaussian distribution with a very narrow kernel, and it enhances the maximum value which makes tree detection easier (Gebreslasie et al. 2011). The Gaussian filter is calculated based on Eq. 6.1 (Nixon and Aguado 2002):

$$G(x, y) = \frac{1}{2\pi\sigma^2} e^{-\frac{x^2+y^2}{2\sigma^2}} \tag{6.1}$$

where σ is the standard deviation of the distribution and was set to 3. The kernel size represents the size of neighborhood where the maximum value is located, and it should be assigned corresponding to expected crown size and flight altitude. The higher the flight altitude is, the smaller the kernel size will be. Figure 6.3 shows an intersect line on CHM Gaussian filtered image plotted and corresponding RGB image. After CHM noise reduction, LM filtering is conducted, where the algorithm finds the brightness pixels within the same kernel size. Figure 6.4 shows the results of LM filtering on smoothed CHM in three plots of 9, 13, and 18 with one sample line intersect in each plot.

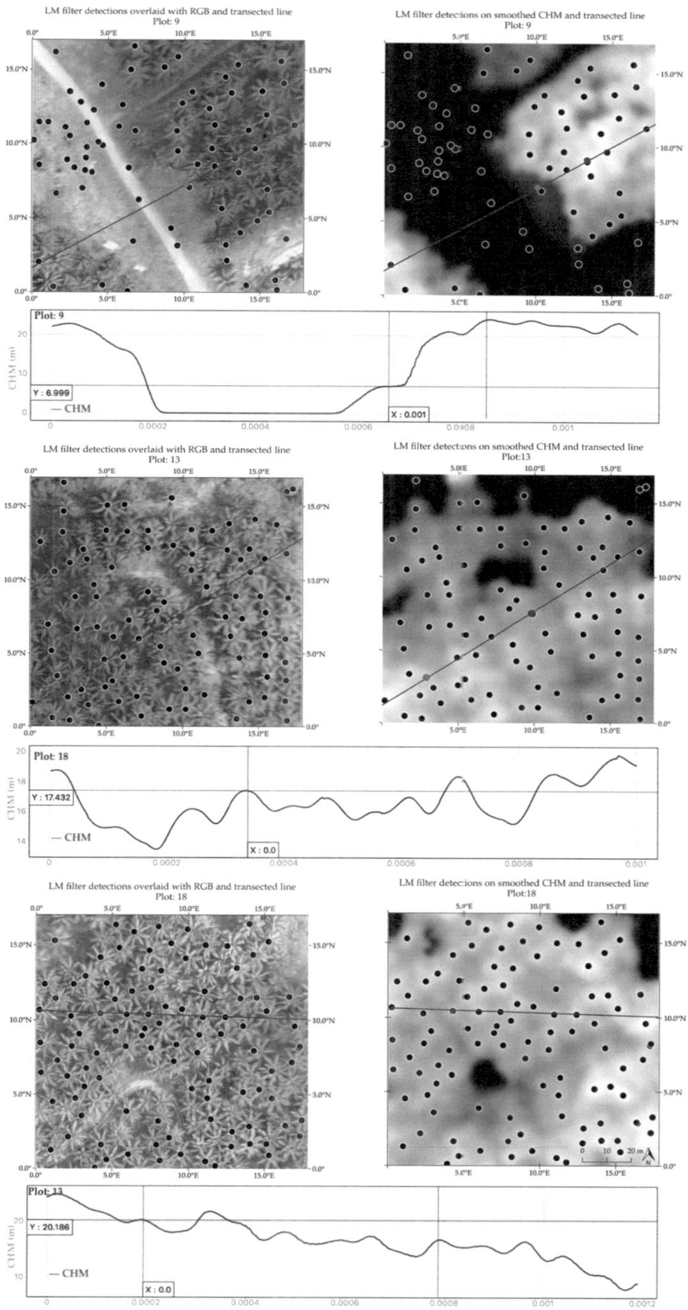

Fig. 6.4 LM filtered detections on smoothed CHM with line transect in three plots of 9, 13, and 18. A-1, B-1, and C-1 show LM filtered detections on smoothed CHM. A-2, B-2, and C-2 show detections overlaid on RGB image. A-3, B-3, and C-3 show CHM transect where perpendicular blue line shows true positive detection and perpendicular red line shows false detection

6.2.2 *Application of Template Matching Algorithm for Individual Tree Detection*

Template in TM approach is a sub-image of a larger image where the TM algorithm tries to find occurrence of this template in the large image based on the template's boundary. Among several TM techniques, normalized cross correlation (NCC) and square root of sum of squared differences (SSD) are commonly used to measure similarity, and other methods such as sum of absolute differences (SAD) and sequential similarity detection algorithm (SSDA) are utilized for pattern recognition and so on (Ahuja and Tuli 2013). NCC is one of the common approaches in TM, and it is used as a measure for degree of agreement between two compared images. Technically NCC algorithm searches the location of the maximum value to determine the matching point between template and image. The NCC of the template t *(2 h + 1),(2w + 1)* and image x at the location of *(u,v)* is calculated as (Ahuja and Tuli 2013):

$$c(u,v) = \frac{\sum_{i=-h}^{h} \sum_{j=-w}^{w} X(i,j)T(i,J)}{\sqrt{\sum_{i=-h}^{h} \sum_{j=-w}^{w} X(i,j)^2} \sqrt{\sum_{i=-h}^{h} \sum_{j=-w}^{w} T(i,j)^2}} \tag{6.2}$$

with

$$X(i,j) = x(u+i, v+j) - \overline{x}$$
$$T(i,j) = t(h+i, w+j) - \overline{t}$$

In this chapter, NCC is used to detect individual trees. To improve the quality of the detections, an optimized template should be selected based on an iterative process where the performances of several templates need to be checked and the one with better performance is considered as the main template (Fig. 6.5).

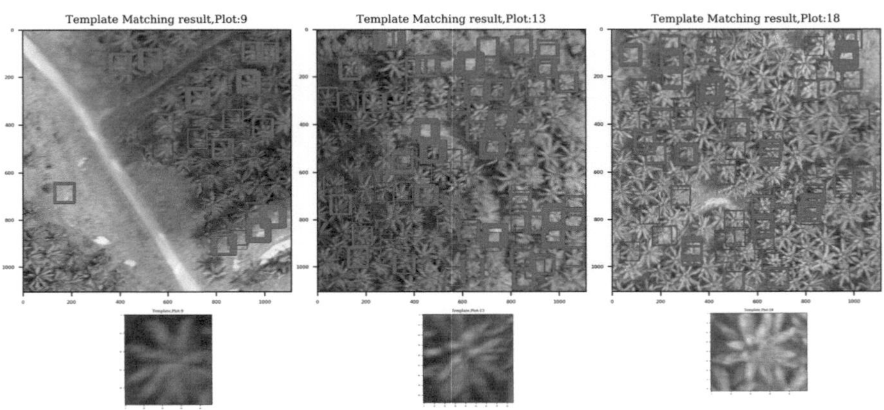

Fig. 6.5 TM detected trees in different plots of the chosen template

6.3 Individual Tree Detection Accuracy Assessment

To check the accuracy of individual tree detection, appropriate number of sample plots is considered. In this chapter we consider 20 randomly selected plots of 200 m^2 (100 m × 100 m), and a comparison between LM-based and TM-based tree detection via visual assessment on UAV image is conducted. The accuracy in each of these plots is evaluated in terms of true positive (TP; correct oil palm detection), false negative (FN; correspond to omission errors), false positive (FP; correspond to commission errors), recall, precision, and F-score using Eqs. 6.3, 6.4 and 6.5 (Li et al. 2012);

$$\text{Precision} = \frac{TP}{TP + FP} \tag{6.3}$$

$$\text{Recall} = \frac{\text{TP}}{\text{TP} + \text{FN}} \tag{6.4}$$

$$\text{F} - \text{measure} = \frac{2 \times \text{Recall} \times \text{Precision}}{\text{Recall} + \text{Precision}} \tag{6.5}$$

In this context, precision can be interpreted as the probability that a detected oil palm tree has correctly been detected, and recall is a measure of trees detected and is inversely related to commission error. The F-score is defined as the harmonic mean of precision and recall, where it can be used as an overall performance metric.

Table 6.1 shows precision, recall, and F-score for each randomly selected plot.

The overall precision values of 0.90 and 0.72, respectively, for LM-CHM and TM-RGB methods are ranged between 0.49 to 1.00 and 0.53 to 0.89. The overall recall values for LM-CHM and TM-RGB are 0.78 and 0.53 ranging between 0.64 to 0.87 and 0.21 to 0.91. The F-score for LM-CHM method is higher than TM-RGB methods which, respectively, are 0.83 and 0.60. From 17,252 oil palm trees in the 20 plots, LM algorithm could detect 1395 (almost 80%) and TM algorithm, 967 (almost 55%). Three hundred fifty-seven and 785 trees have been missed, respectively, in LM-CHM and TM-RGB approaches. In both cases, background vegetation incorrectly has been labeled as oil palm tree, where 141 and 322 objects have been falsely detected as oil palm tree in LM-CHM and TM-RGB approaches. LM-CHM worked better in almost all of the plots; however the performance decreased in highly dense area. Contradictory to TM-RGB approach, in LM-CHM approach, shadows did not affect the performance as it was reflected in precision value. In densely cultivated plots due to leaves overlapping of neighboring trees, algorithm failed, and also in sparsely cultivated plots, shadows caused some commission errors (FP). Inherited distortion of UAV image also caused some omission errors (FN) in TM-RGB approach.

Table 6.1 Accuracy of LM-CHM and TM-RGB approaches in 20 randomly selected plots

	LM-CHM			TM-RGB		
Plot ID.	Precision	Recall	F-score	Precision	Recall	F-score
1	0.96	0.74	0.84	0.65	0.32	0.43
2	0.88	0.81	0.84	0.79	0.74	0.76
3	0.82	0.74	0.78	0.82	0.38	0.52
5	0.95	0.81	0.87	0.89	0.48	0.62
6	0.91	0.87	0.89	0.74	0.61	0.67
9	0.49	0.65	0.56	0.67	0.42	0.52
10	0.95	0.81	0.88	0.64	0.33	0.43
11	0.97	0.86	0.91	0.81	0.55	0.65
12	0.98	0.83	0.90	0.65	0.28	0.39
13	0.96	0.81	0.87	0.78	0.76	0.77
14	0.95	0.64	0.77	0.53	0.29	0.37
15	0.57	0.67	0.62	0.62	0.21	0.32
16	0.87	0.68	0.76	0.62	0.41	0.49
17	0.97	0.85	0.90	0.85	0.72	0.78
18	0.99	0.86	0.92	0.88	0.79	0.83
19	0.98	0.84	0.91	0.73	0.49	0.59
20	0.95	0.78	0.86	0.87	0.83	0.85
21	0.98	0.82	0.89	0.59	0.36	0.45
22	1	0.73	0.85	0.66	0.87	0.75
23	0.92	0.84	0.88	0.77	0.91	0.83
Avg.	0.90	0.78	0.83	0.72	0.53	0.60

6.4 Individual Oil Palm Height Estimation

As the LM-CHM showed better performance in individual tree detection in this particular example, detected trees from this approach are taken for height estimation. In general, in highly suitable forest area, oil palm height may reach to 30 m. However, due to yield reduction and difficulty in harvesting, normally when tree height exceeds 15–18 m, it will be replanted (Corley and Tinker 2003).

To make sure about accuracy of UAV-derived information, the consistency of UAV-derived DSM was compared with Advanced World 3D (AW3D) DSM at the locations of the detected trees (Fig. 6.6). AW3D in 5 m spatial resolution generated based on Panchromatic Remote-sensing Instrument for Stereo Mapping (PRISM) sensor operation on board the Advanced Land Observing Satellite (ALOS) (Takaku et al. 2016). Based on the comparison presented in Fig. 6.5, the geolocation errors are not significant, and UAV-DSM is consistent over the whole area, and also strong correlation was observed between two DSMs.

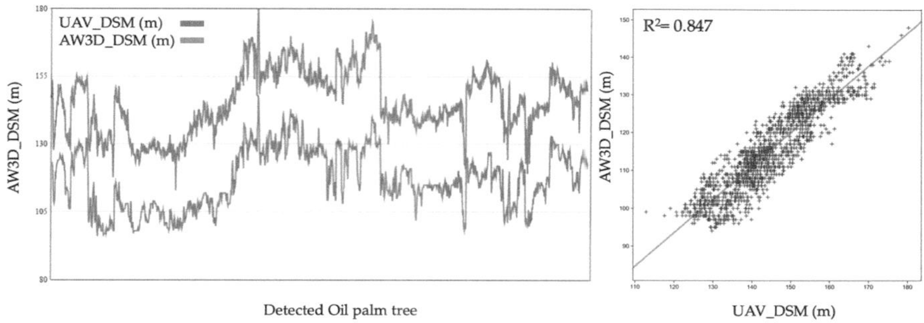

Fig. 6.6 Comparison of UAV-derived DSM and AW3D-derived DSM

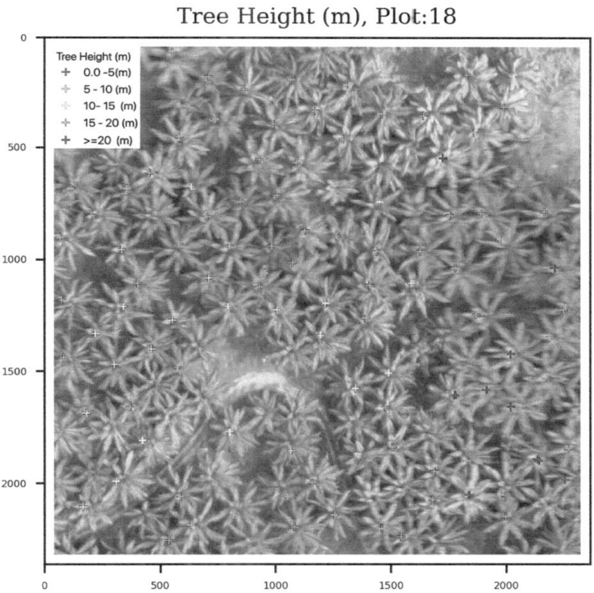

Fig. 6.7 Detected oil palm associated with their estimated heights

Extracting CHM values at the location of detected trees, individual oil palm height information is generated (Fig. 6.7). Figure 6.8 shows tree height distributions, in which, in all sample plots, tree height is 15 m or higher. Plots 2, 3, 9, 15, and 16 showed greater false-positive errors where the background vegetations were incorrectly detected as oil palm tree.

According to the result, replanting plan should be considered for the whole site; however, plots 1, 12, and 16 can be prioritized for replanting plan as they have the tallest trees among all the sample plots.

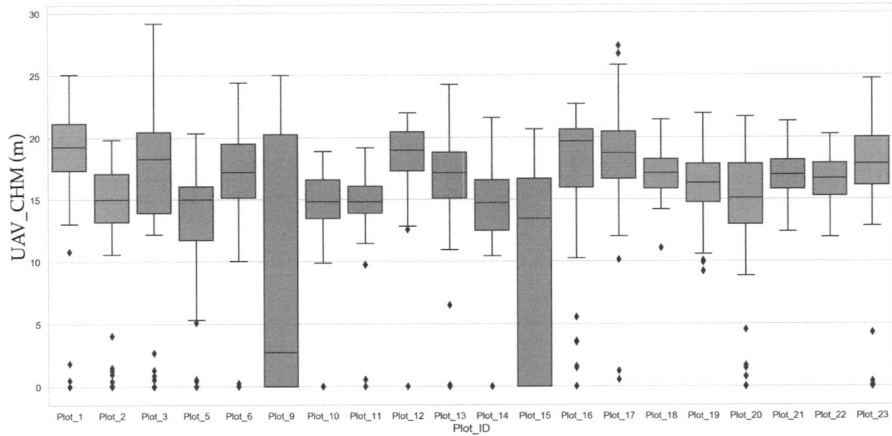

Fig. 6.8 Oil palm height distribution in the sample plots

6.5 Summary

In this chapter, we introduced a general work flow for oil palm tree detection and height measurements using UAV. We assessed two commonly used algorithms of LM and TM. LM algorithm applied on UAV-derived CHM showed better performance for individual tree detection in this case study. However UAV-derived CHM proved to be useful for oil palm height measurements, but individual tree detection from optical sensors still needs much improvement (Koh and Wich 2012), particularly in dense plantations where crown overlapping between neighboring trees is observed. Also crowns shadows may affect the performance of the algorithms. Using different deep learning models along with LiDAR-based imagery has been reported to show better performance in individual tree detection (Malek et al. 2014; Li et al. 2016; Mohan et al. 2017); however, high cost of LiDAR sensors is counted as one of the important barriers in implementation of this technology in practice.

Nonetheless, the success of individual tree detection, regardless of onboard sensor and UAV image processing, is influenced by weather condition, structure of plantation, and user expertise (Eriksson et al. 2011; Dandois et al. 2015; Mohan et al. 2017).

Individual tree detection and tree height measurement are only two applications of UAV in effective oil palm managements, whereas UAV imagery can be used for terrain correction operation by providing 3D modeling of elevation, designing blueprints for replanting operations, etc.

References

AgiSoft, L.L.C. PhotoScan Professional Edition. https://www.agisoft.com
Ahuja K, Tuli P (2013) Object recognition by template matching using correlations and phase angle method. Int J Adv Res Comput
Chemura A, Van Duren I, Van Leeuwen LM (2015) Determination of the age of oil palm from crown projection area detected fromWorldView-2 multispectral remote sensing data: the case of Ejisu-Juaben District, Ghana. ISPRS J Photogramm Remote Sens 100:118–127. https://doi.org/10.1016/j.isprsjprs.2014.07.013
Colomina I, Molina P (2014) Unmanned aerial systems for photogrammetry and remote sensing: a review. ISPRS J Photogramm Remote Sens 92:79–97. https://doi.org/10.1016/j.isprsjprs.2014.02.013
Corley RHV, Tinker PB (2003) The oil palm, 4th edn. Blackwell Science Ltd, Oxford
Dandois JP, Olano M, Ellis EC (2015) Optimal altitude, overlap, and weather conditions for computer vision Uav estimates of Forest structure. Remote Sens 7(10):13895–13920. https://doi.org/10.3390/rs71013895
Dow JM, Neilan RE, Rizos C (2009) The international GNSS service in a changing landscape of global navigation satellite systems. J Geod 83(3–4):191–198. https://doi.org/10.1007/s00190-008-0300-3
Eriksson M, Gougeon FA, Brandtberg T, Descombes X, Larsen M, Perrin G (2011) Comparison of six individual tree crown detection algorithms evaluated under varying Forest conditions. Int J Remote Sens 32(20):5827–5852. https://doi.org/10.1080/01431161.2010.507790
Gaveau DLA, Sheil D, Husnayaen, Salim MA, Arjasakusuma S, Ancrenaz M, Pacheco P, Meijaard E (2016) Rapid conversions and avoided deforestation: examining four decades of industrial plantation expansion in Borneo. Sci Rep (October 2015) Nature Publishing Group 6:1–13. https://doi.org/10.1038/srep32017
Gebreslasie MT, Ahmed FB, Van Aardt JAN, Blakeway F (2011) Individual tree detection based on variable and fixed window size local maxima filtering applied to IKONOS imagery for even-aged Eucalyptus plantation forests. Int J Remote Sens 32(15):4141–4154. https://doi.org/10.1080/01431161003777205
Gutiérrez-Vélez VH, Defries R (2013) Annual multi-resolution detection of land cover conversion to oil palm in the Peruvian Amazon. Remote Sens Environ 129:154–167. https://doi.org/10.1016/j.rse.2012.10.033
Khai LC, Kasturi DK, Christine P, Kian PT (2017) A review of remote sensing applications for oil palm studies. Geo-Spat Inf Sci 20(2):184–200. https://doi.org/10.1080/10095020.2017.1337317
Koh LP, Wich SA (2012) Dawn of drone ecology: low-cost autonomous aerial vehicles for conservation. Trop Conserv Sci 5(2):121–132. https://doi.org/10.1177/194008291200500202
Li W, Guo Q, Jakubowski MK, Kelly M (2012) A new method for segmenting individual trees from the Lidar point cloud. Photogramm Eng Remote Sens 78(1):75–84. https://doi.org/10.14358/pers.78.1.75
Li W, Fu H, Yu L, Cracknell A (2016) Deep learning based oil palm tree detection and counting for high-resolution remote sensing images. Remote Sens 9(1):22. https://doi.org/10.3390/rs9010022
Maes WH, Steppe K (2019) Perspectives for remote sensing with unmanned aerial vehicles in precision agriculture. Trends Plant Sci Elsevier Ltd 24(2):152–164. https://doi.org/10.1016/j.tplants.2018.11.007
Malek, Salim, Yakoub Bazi, Naif Alajlan, Haikel AlHichri, and Farid Melgani. 2014. "Efficient framework for palm tree detection in UAV images." IEEE J Sel Top Appl Earth Obs Remote Sens 7 (12): 4692–4703. IEEE doi:https://doi.org/10.1109/JSTARS.2014.2331425

Mohan M, Silva CA, Klauberg C, Jat P, Catts G, Cardil A, Hudak AT, Dia M (2017) Individual tree detection from unmanned aerial vehicle (UAV) derived canopy height model in an open canopy mixed conifer forest. Forests 8(9):1–17. https://doi.org/10.3390/f8090340

Nixon, MS, Aguado AS (2002) Feature extraction and image processing. First. Newnes an imprint of Butterworth-Heinemann. https://doi.org/10.1016/b978-0-12-396549-3.00003-3

Novotný J, Hanuš J, Lukeš P, Kaplan V (2011) Individual tree crowns delineation using local maxima approach and seeded region growing technique. GIS Ostrava 26:1. http://gisak.vsb.cz/GIS_Ostrava/GIS_Ova_2011/sbornik/papers/Novotny.pdf

Park JS, Kim J-I, La PH, Ye SL, Wook Pyeon M, Mi HL (2016) Calculation of tree height and canopy crown from drone images using segmentation. J Korean Soc Surv Geod Photogramm Cartogr 33(6):605–614. https://doi.org/10.7848/ksgpc.2015.33.6.605

Pouliot DA, King DJ, Bell FW, Pitt DG (2002) Automated tree crown detection and delineation in high-resolution digital camera imagery of coniferous Forest regeneration. Remote Sens Environ 82:322–334. http://ac.els-cdn.com/S0034425702000500/1-s2.0-S0034425702000500-main.pdf?_tid=12e9edda-4c0b-11e7-b0ee-00000aacb35d&acdnat=1496899790_00ad3e6fe518e7b46b112c38ca28c3ed

RSPO (2013) RSPO principles and criteria for sustainable palm oil production. https://rspo.org/keydocuments/certification/rspo-principles-and-criteria

Santoso H, Gunawan T, Jatmiko RH, Darmosarkoro W, Minasny B (2011) Mapping and identifying basal stem rot disease in oil palms in North Sumatra with QuickBird imagery. Precis Agric 12:233–248. https://doi.org/10.1007/s11119-010-9172-7

Takaku J, Tadono T, Tsutsui K, Ichikawa M (2016) Validation of 'AW3D' global DSM generated from ALOS prism. ISPRS Ann Photogramm Remote Sens Spat Inf Sci 3(July):25–31. https://doi.org/10.5194/isprs-annals-III-4-25-2016

Tan KP, Kanniah KD, Cracknell AP (2013) Use of UK-DMC 2 and ALOS PALSAR for studying the age of oil palm trees in southern peninsular Malaysia. Int J Remote Sens 34(20):7424–7446. https://doi.org/10.1080/01431161.2013.822601

Tian J, Wang L, Li X, Gong H, Shi C, Zhong R, Liu X (2017) Comparison of UAV and WorldView-2 imagery for mapping leaf area index of Mangrove Forest. Int J Appl Earth Obs Geoinf 61:22–31. https://doi.org/10.1016/j.jag.2017.05.002

USDA (2019) Oilseeds: World markets and trades. https://apps.fas.usda.gov/psdonline/circulars/oilseeds.pdf

Wallace L, Lucieer A, Watson CS (2014) Evaluating tree detection and segmentation routines on very high resolution UAV LiDAR Ata. IEEE Trans Geosci Remote Sens 52(12):7619–7628. https://doi.org/10.1109/TGRS.2014.2315649

Westoby MJ, Brasington J, Glasser NF, Hambrey MJ, Reynolds JM (2012) 'Structure-from-motion' photogrammetry: a low-cost, effective tool for geoscience applications. Geomorphology Elsevier BV 179:300–314. https://doi.org/10.1016/j.geomorph.2012.08.021

Woittiez LS, Van Wijk MT, Slingerland M, Van Noordwijk M, Giller KE (2017) Yield gaps in oil palm: a quantitative review of contributing factors. Eur J Agron 83:57–77. https://doi.org/10.1016/j.eja.2016.11.002

Wulder M, Niemann KO, Goodenough DG (2000) Local maximum filtering for the extraction of tree locations and basal area from high spatial resolution imagery. Remote Sens Environ 73:103–114. http://ac.els-cdn.com/S0034425700001012/1-s2.0-S0034425700001012-main.pdf?_tid=4f55b026-83d6-11e7-a385-00000aab0f6b&acdnat=1503034395_c6fd49bf72d3645837f424d2bf84782f

Chapter 7
Applications of UAVs in Plantation Health and Area Management in Malaysia

Ram Avtar, Stanley Anak Suab, Ali P. Yunus, Pankaj Kumar, Prashant K. Srivastava, Manish Ramaiah, and Churchill Anak Juan

Abstract The scope of unmanned aerial vehicles (UAVs), also known as "drone technology," is increasing, with various applications in the field of remote sensing and environment. UAVs not only provide high-resolution, real-time data, but also have different applications for end users. They have become an essential tool for land surveyors because traditional land survey methods are expensive and time-consuming, requiring trained professionals and many hours to measure a single plot of land. With the advancement of UAVs, we can significantly reduce the cost. In this study, we have collected UAV data in Malaysia to acquire information about the plantation management practices, as well as oil palm health assessment. Our results showed that multispectral data collected from a UAV-borne MicaSense RedEdge camera is useful for identifying physiological stress in mature oil palm plants, which clearly illustrates stunted tree crown with low value of normalized difference vegetation index (NDVI).

Keywords VARI · Plant stress · NDVI · UAVs · Rhinoceros beetles

---

R. Avtar (✉) · S. A. Suab · M. Ramaiah
Faculty of Environmental Earth Science, Hokkaido University, Sapporo, Hokkaido, Japan
e-mail: ram@ees.hokudai.ac.jp

A. P. Yunus
State Key Laboratory of Geohazard Prevention and Geo-environment Protection, Chengdu University of Technology, Chengdu, Sichuan, China

P. Kumar
Natural Resources and Ecosystem Services, Institute for Global Environmental Strategies, Hayama, Japan

P. K. Srivastava
Institute of Environment and Sustainable Development, Banaras Hindu University, Varanasi, India

C. A. Juan
CGIS Enterprise, Kuching, Sarawak, Malaysia

R. Avtar, T. Watanabe (eds.), *Unmanned Aerial Vehicle: Applications in Agriculture and Environment*, https://doi.org/10.1007/978-3-030-27157-2_7

7.1 Introduction

Remotely piloted aerial systems (RPAS) or unmanned aerial vehicles (UAVs) have various applications, with one of the common fields ~~is~~ being agriculture and forestry. There is an increasing interest ~~of~~ in high-precision UAV techniques as high-resolution data can be collected flexibly, quickly, and at a relatively low price. Also, drones play an important role in filling the gaps in data collected using manned aircraft or satellite remote sensing. Moreover, they have many advantages in research and development fields, particularly in land development projects (Banu et al. 2016). Recently, UAV data collection is popular, but processing and extracting useful information is under development. Recent developments ~~of~~ in operational small UAVs can provide high spatiotemporal ortho-image compared to conventional remote sensing data. UAV data is useful ~~to~~ in providing three-dimensional (3D) structure of objects using structure from motion (SfM) techniques that utilize 3D point cloud data (Guerra-Hernández et al. 2016).

In the last decade, remote sensing techniques applied to agriculture and forestry have been receiving increased attention, which leads to enable the extraction of important information for planning and sustainable resource management (Shao 2012). When relying on satellite images of plantation areas, there is a substantial lag in terms of acquisition. There have been various studies that show the potential of satellite data in land use/land cover (LULC) monitoring, crop monitoring, forest monitoring, etc. (Avtar et al. 2012, 2017; Breckenridge et al. 2012; Li et al. 2014; Kalantar et al. 2017). However, the major limitations of currently available satellite data are cost, spatial resolution, clouds, detailed information about objects, etc. (Attarzadeh and Momeni 2012; Gevaert et al. 2015). Low-altitude UAV is a promising and affordable technology for precision agriculture to acquire detailed information about objects (Shamshiri et al. 2018; Ballesteros et al. 2014). The global market for agricultural UAV drones is estimated to reach 3.7 billion US dollars by the year 2022 (Source: Radiant Insight Research firm). With the advancement of UAV technology and processing algorithms, there are various applications. For example, Luna and Lobo (2016) used UAVs for assessing crop quality by mapping crop canopy gaps and using this information for replanting. UAV data is not only useful in field surveys but also for topographic analysis based on photogrammetry (Jaud et al. 2016). This information can be useful to understand the local topography. Tattaris et al. (2016) compared three remote sensing approaches using a UAV, proximal sensing, and satellite-based imagery for canopy temperature and NDVI measurement. They concluded that UAV-based techniques are best for measuring canopy temperature and NDVI. Therefore, UAVs have great potential to monitor various vegetation parameters and phenological changes. UAVs can carry various sensors such as digital cameras, multispectral cameras, hyperspectral sensors, infrared thermal imagers, light detection and ranging (LiDAR), etc. The selection of the sensor depends on the objectives of one's study. Sensors can provide live information about objects. Such information contributes to the in-depth analysis of the objects. UAV techniques can provide us accurate area information about the plants, area for replanting or thinning, making inventory, finding distances between plants, calculating canopy diameter, plant height, plant density, as well as creating digital surface model (DSM), digital terrain model (DTM), NDVI, yield, biomass, health and stress, etc. (Shamshiri et al. 2018). Multispectral and hyperspectral sensors attached to UAVs are popular to assess vegetation health (Shang et al. 2015; Pullanagari et al. 2016).

Satellite programs like Landsat are still one of the most commonly used satellites worldwide in the field of forestry, agriculture, land use/land cover, etc. (Alberts 2012). Features of the recent Sentinel-2 include higher revisit frequency, narrower bandwidths, and finer resolution (Hojas-Gascón et al. 2015). However, it is still limited and not useful to study individual trees. From this perspective, the applicability and effectiveness of drones can fill the gaps of other data types to a large extent even if current forest applications are still in the experimental stage (Shahbazi et al. 2014). Initially, UAVs were used for military or surveillance purposes, but due to advancements in technology/availability/cost, they have a wide spectrum of applications in the field of precision agriculture, forestry, biodiversity, meteorology, wildlife research, land management, etc. (Shahbazi et al. 2014). Uses of UAVs include estimation of tree crown diameters, tree fall, and canopy openness (Inoue et al. 2014), detection of nutrient and water deficiencies in agricultural crops (Martínez et al. 2017), and assistance in the management of insect pests (Severtson et al. 2016). This study shows how UAV data can be used in plantation land management to identify suitable sites for plantation, as well as monitoring plant health, and furthermore how the precise information on landscapes can reduce the cost of plantation area development.

7.2 Application of UAVs in Plantation Area Development

Current methodologies for managing landscapes often rely on field survey methods and documenting the information, which are labor-intensive and time-consuming processes. If a landscape doesn't have any existing information, then it becomes more tedious to collect the information required for the landscape management plan. Landscape assessment is one promising application of UAVs because it can provide precise information about various objects in the particular landscape, as well as topographic information. Integrated use of existing landscape information with UAV data can contribute to improve efficiency in landscape management. Stek (2016) reported that UAVs can be efficiently used as a source of information in landscape archaeological projects and that UAVs can impact significantly without the need for disproportional investment in terms of time, energy, and other resources. Navarro et al. (2019) found that UAV and Sentinel data can be useful for mangrove plantation and biomass estimation in Senegal. Almeida et al. (2019) used UAV-borne LiDAR system for monitoring structural features in plantation area restoration in Brazil's Atlantic Forest. Fawcett et al. (2019) demonstrated that UAV-based SfM can provide reliable data not only for oil palm inventory generation, but also allows for the retrieval of basic structural parameters which may enable per-palm aboveground biomass estimations. High-resolution UAV data can be useful to know precisely the area covered by various land structures and to avoid invasion or encroachment.

In the field of plantation area development, if there is limited information about the landscape, then use of UAV data can play a significant role by providing high-resolution landscape and topographic information. This information is useful in planning for unplanned areas, for example, planning for plantation by providing information about development of terraces, the distance between the terraces, length of the terraces, labor needed for development of terraces, seedlings required for planting particular species, and business planning. Landscape planning for plantation area development using UAV data can calculate the cost of area development and future benefits.

7.3 Case Studies

7.3.1 UAVs for Plantation Establishment and Replanting

The aerial surveys for the collection of high-resolution aerial photos were conducted on 19 January 2017. The aerial photos were processed with Agisoft software to generate the ortho-mosaic image of the area (Fig. 7.1). Visual interpretation followed by digital interpretation was used to delineate various parameters.

Figure 7.1 shows the area of oil palm plantation in Malaysia. Some of the area is already converted into oil palm plantation, while another portion is under plantation establishment. UAV-captured data can be used to a significant extent for terrace planning for oil palm plantation. The total cost of developing the area and number of seedlings for developing the plantation can also be calculated. We can also estimate the time and money required for the oil palm area development based on ortho-image and topographic information because high elevated areas are labor-intensive and have higher transportation cost compared to low elevated areas. This exercise can be useful for development of a business model to calculate the costs and benefits of starting a project at a particular site by providing more precise information about the area, type of land cover, and topographic information. This case study can also be used to identify problems in the existing oil palm plantation area. Figure 7.2a shows waterlogged area which can impact the yield of oil palm plantation. Waterlogging information is useful for predicting the yield because previous studies show that oil palm yield decreases due to waterlogging (Henson et al. 2008). Therefore, we can avoid areas with probability of waterlogging in the future plantation development plan. Plantation areas with remote access can also be identified using UAV data with topographic information. The probability of management of remote plantation areas is poor and labor-intensive. Therefore, during the development of plantation areas, we can consider either present remote locations or find areas of better accessibility to workers. Ortho-image also shows oil palm vigor to identify

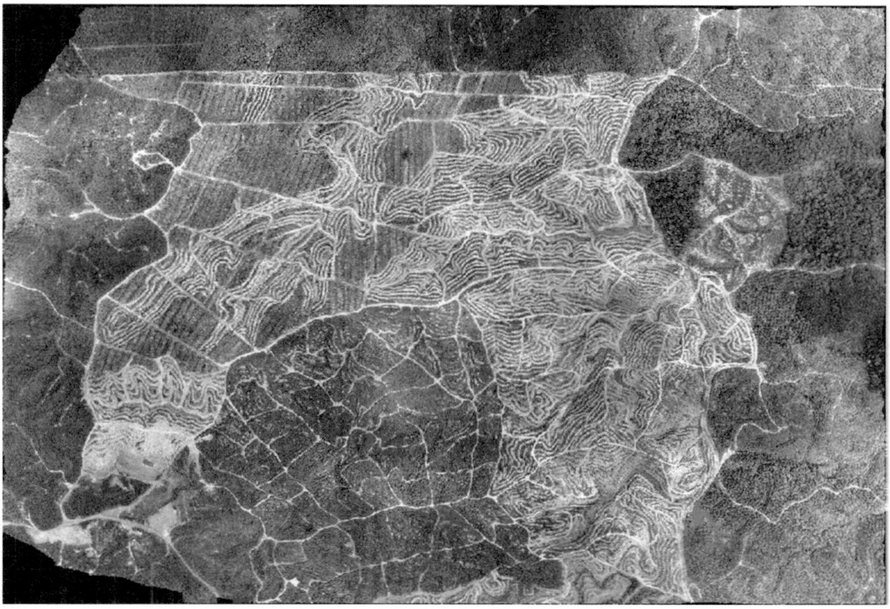

Fig. 7.1 Ortho-mosaic image of the plantation area development in Malaysia

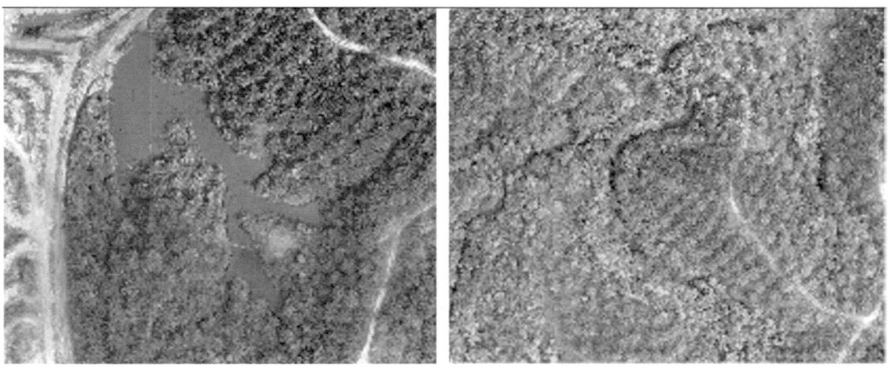

Fig. 7.2 (**a**) Waterlogged oil palm, (**b**) oil palm vigor distribution

healthiness of oil palm plantation (Fig. 7.2b). Oil palm vigor is an important parameter for predicting the yield as well as identifying areas that need additional fertilizer for growth. The use of ortho-image of oil palm plantation can be useful to identify areas with stunted growth, unhealthy oil palm, waterlogged areas, inaccessible areas with unreachable palms, etc.

Oil palm plantation areas are designed and divided into blocks due to economical and practical reasons (Chong et al. 2017). In flat areas, blocks are designed as same

Fig. 7.3 Oil palm replanting block planning on flat land (ortho-mosaic and DEM)

size with square shape, while in undulated or hilly terrain, the blocks' size and shapes are uneven following natural features as boundaries. Some hilly terrains need terracing to maximize planting density and as countermeasure for surface water runoff (Shafri et al. 2012). During land preparation for establishment or replanting, residual plants and old palm trees were piled in rows to decompose. Triangular planting pattern with 9 m distance normally used for flat areas is said to provide maximum penetration of sunlight (Basiron 2007). Using three-dimensional (3D) digital elevation model (DEM) generated from aerial photos is useful for users to identify the type of block design needed for a specific site. Figure 7.3 shows an example of flat terrain with the location of young and old oil palm plants and sites newly prepared for the replanting. Site preparation work progress can be monitored and carefully planned using high-resolution aerial photos collected by UAV.

Figure 7.4 shows the oil palm plantation development in hilly terrain by terracing contour using ortho-mosaic and DEM data. In case of plantation area development in hilly terrain, high-resolution DEM data significantly plays a major role in designing the contours for plantation.

Previous practice involved manual fieldwork to a large extent using handheld GPS to measure and check the terraces. It was labor-intensive and time-consuming. UAV-acquired data can help in estimating the block level overall length of terraces in Geographic Information Systems (GIS). This process can help by saving time and manual work. A semiautomated generation of contours helps the GIS operators to validate the work quality of site preparations. Most importantly, payment to the contractor is based on the amount of site preparation work done. Figure 7.5 shows an example of terraces for rubber plantation on a hilly area.

Fig. 7.4 Hill terraces for new oil palm plantation (ortho-mosaic and DEM)

Fig. 7.5 Hill terraces of rubber plantations in 3D

7.3.2 UAVs for Plantation Management and Maintenance

The UAV technology is helpful in effective plantation management and maintenance work, especially for oil palm where intensive care is needed at all stages of crop development. Weed control is very important at the early growth stage of oil palm plantation because it can compete by depriving the young sapling of water, nutrients, and sunlight resulting in its stunted growth. Some weed species also serve as hosts for diseases/pest/insects. Manual removal of weeds from young plantation sites is time-consuming, costly, and impractical (www.plantationsolutions.com). There are various methods to control the weed growth in the plantation area, namely, cultural, mechanical, and integrated control. The use of livestock and herbicides is the most common

Fig. 7.6 Newly planted oil palm blocks on undulated land

practice in oil palm plantations (Mohamad et al. 2010). High-resolution UAVs' aerial photos can provide information about the condition of the blocks with weeds. Based on the weed area coverage, we can calculate the quantity of herbicides required for treatment as well as post-treatment effect. Herbicide treatments with poor efficacy caused weeds to grow and recover faster. Therefore, we can effectively use multi-temporal UAV data to see the effect of herbicides. Figure 7.6 shows the location of various blocks with young oil palm. UAV data can also be useful for designing various structures such as culverts and short bridges for movement of water and vehicles, respectively. These culverts are installed with the appropriate openings to facilitate the flow of surface water away from the road (Liang 2008).

Tree counting is important in plantation management and maintenance as it can provide useful information about the current status and conditions of the plantation area. Tree counting using conventional methods is time-consuming and costly. Furthermore, there is high possibility of human error. Estimation of total tree counts in a plantation area by simply multiplying the total area and number of trees per hectare can lead to either under- or overestimation without considering geographic features such as topography, terrain, forest, or rivers. In addition, the losses and cost of treating affected plants can be correctly estimated by knowing the exact area of affected plants (Chong et al. 2017). Tree counting is also important and necessary for

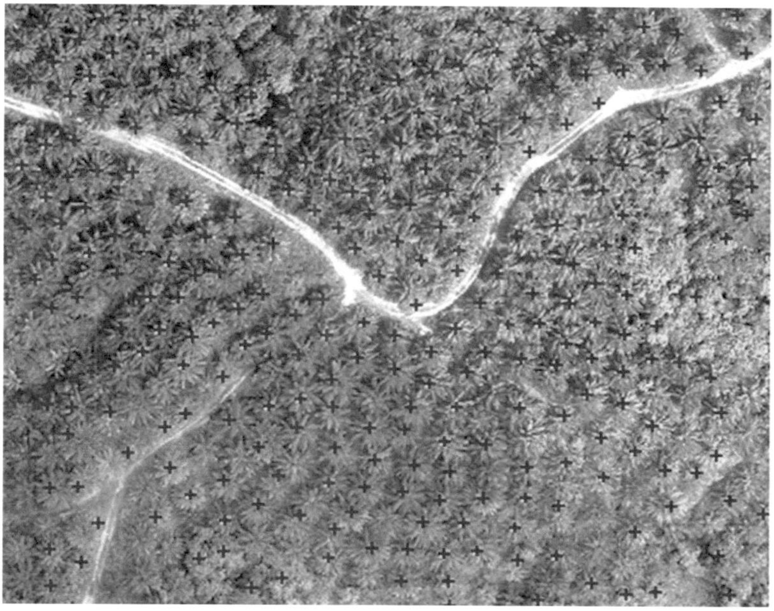

Fig. 7.7 Oil palm tree identification using UAV data

yield estimation, fertilizer estimation, harvesting, planning, and managing the growing situation of the palm trees to maximize their productivity. Tree counting is mostly done manually using visual interpretation in GIS to achieve a high level of accuracy, although recently there are various algorithms to automatically identify trees. Wang et al. (2019) used UAV data to identify oil palm trees using the histogram of oriented gradient (HOG) approach with 99% accuracy. Li et al. (2019) used a two-stage convolution neural network to detect individual oil palm trees using QuickBird data. Figure 7.7 shows oil palm tree identification using object-based classification of UAV data. Use of UAV data can be useful for updating tree inventory every year with low cost. UAV-based individual tree identification can also help to provide specific treatment to infected trees and reduce further disease or pest infestation.

7.4 Plantation Health Assessment

Health assessment of oil palm plantation areas is vital for spotting pest and fungal infection as well as bacterial disease. Due to extensive stretches of oil palm plantation and topographic variations, the early detection of these infections is a significant challenge. Traditionally, the human eye has been used to visually identify disease with some training or experience in detecting plant disorders. Using our eyes, the detection of disease can only be recognized at the late stage with some bias and human errors (Bock et al. 2010). UAV data can be used to capture the ortho-image of plantation areas using visible RGB camera, near infrared (NIR), hyperspectral, multispectral sensors, and thermal camera (Ahmed et al. 2019). Variation in temporal, spatial reflectance, and thermal properties can be identified before they can be detected by the human eye and can be associated with palm health for early response (Shamshiri et al. 2018). Examples include the ability of NDVI cameras to calculate the vegetation index describing the relative density and health of the oil palm and of thermal cameras in showing the heat signature of plants (Shamshiri et al. 2018).

7.4.1 Oil Palm Health Monitoring

In a case study, we have captured UAV data in Sandakan, Sabah, using RGB camera. Figure 7.8a shows the ortho-image of the study area. The ortho-image clearly shows variations in the vigor of oil palm. Visible Atmospherically Resistant Index (VARI) was used to see the RGB index for leaf coverage. VARI has been used by other studies that indicate its usefulness in monitoring vegetation health (Song et al. 2015; Hou et al. 2016). VARI $= R_{Green} - R_{Red}/R_{Green} + R_{Red} - R_{Blue}$.

VARI is a useful parameter to identify the health of oil palm. Figure 7.8b shows the variations in the VARI of the study area. Gray or yellow color represents stressed or unhealthy oil palm, whereas green represents healthy oil palm plants. In the case of dead oil palm, the value of VARI is very low. Therefore, we can use the VARI to get the information about oil palm health in the study area.

Figure 7.9a shows the location of the vacant points that show dead oil palm trees. In the study area, there are 87 dead trees. The blue points show those locations of the dead trees (Fig. 7.8a). The identification of dead trees is useful in order to take preventive measures for controlling the further spread of insects/diseases in the area. Figure 7.9b shows the total area of stressed oil palm plantation area. An area of 11.86 hectares of the oil palm plantation is under stress. This information is useful for identifying the cause of oil palm stress; and based on this information, preventive measures such as spraying of fertilizers or pesticides can be recommended. However, we need to conduct a ground survey with experts to make a final decision.

7.4.2 Detection of Oil Palm Pests Using MicaSense RedEdge Sensor

Detection of pests and diseases is one of the major applications of UAV for managing oil palm plantations. This is because it can provide information about early detection of pests and diseases in a wider area that could help in planning intervention strategies and preventing outbreak (Chong et al. 2017). *Oryctes*

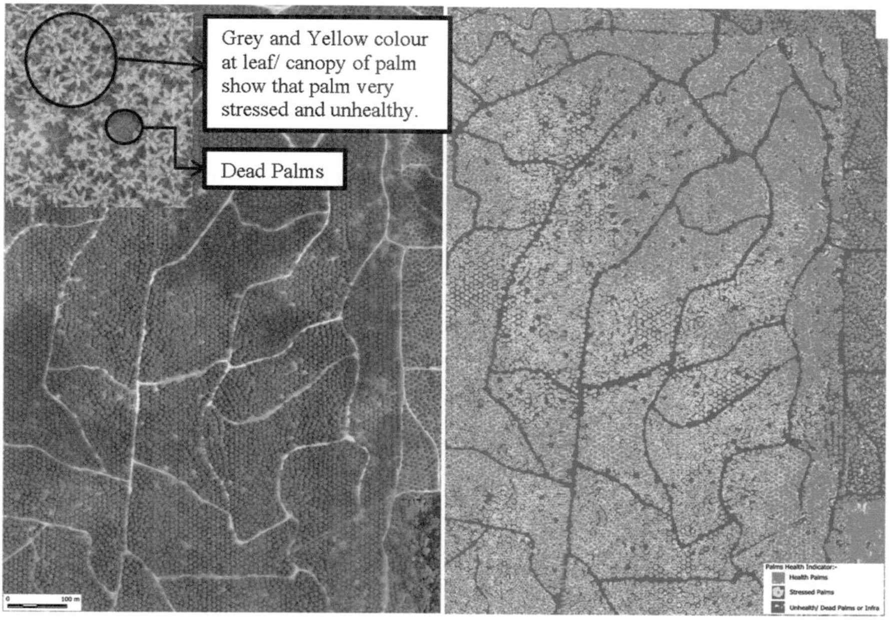

Fig. 7.8 (**a**) UAV-captured ortho-image of plantation area, (**b**) VARI image of oil palm area

rhinoceros beetles and *Ganoderma* are common infections that prevent the growth of oil palm trees and yields (Mostafa et al. 2018). Previous studies indicated that infection caused by rhinoceros beetles can significantly reduce yield by about 25–30% (Liau and Ahmad 1993; Potineni and Saravanan 2013). Advancements in remote sensing technology with UAV applications can be used to diagnose the pest and disease infestation based on the symptoms of plants (Liaghat et al. 2014). Reflectance spectra of vegetation contain information on plant pigment concentration, leaf cellular structure, and leaf moisture content, and this information can be useful to identify the infected plants. In this case study, we used MicaSense RedEdge sensor to study the capability of multispectral imaging to detect the change in oil palm crown infected by rhinoceros beetles. A quick and efficient method of detecting and mapping rhinoceros beetle infection at the field level will assist growers to better manage and control this disease. The method can also benefit growers financially. We have captured UAV data at the flight heights of 20 m, 60 m, and 80 m. Sandino et al. (2018) concluded that higher spatial resolutions lead to better results, since more meaningful information can be extracted when plants are analyzed individually. Therefore, we conducted our analysis using 20 m ortho-image data. The oil palm image data obtained from the DJI Phantom 4 and MicaSense RedEdge sensor is shown in Fig. 7.10.

Figure 7.11 shows the identification of oil palm crown and leaf pattern. Rhinoceros beetles can damage immature oil palms by making holes in the base of the

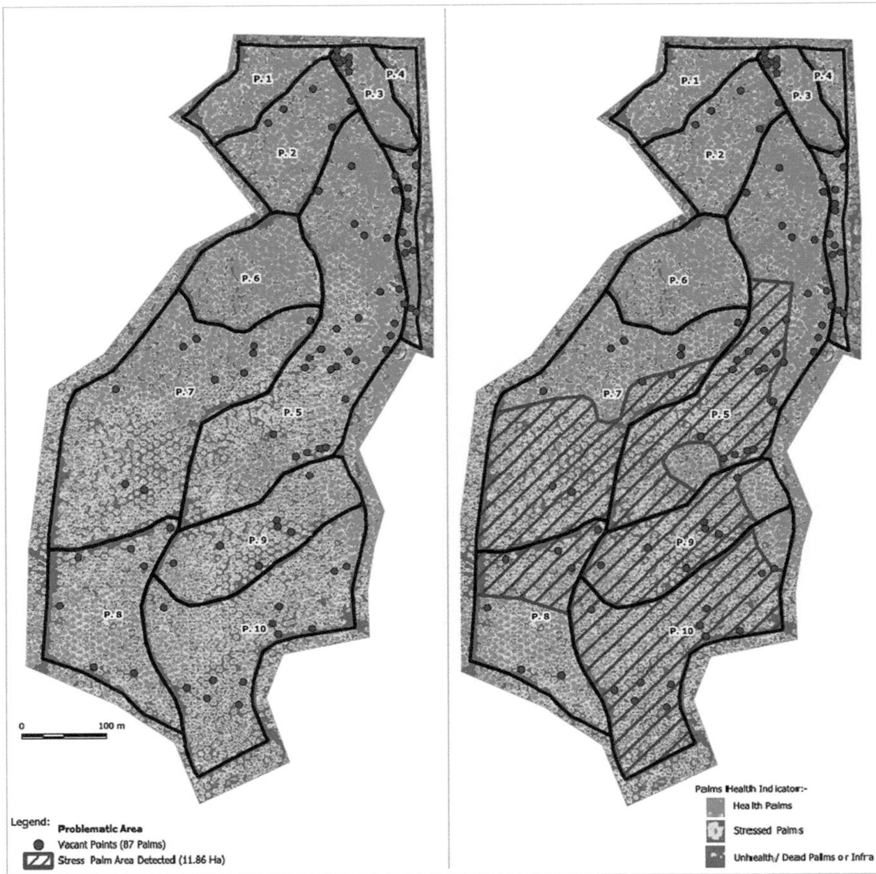

Fig. 7.9 (**a**) Location of the dead trees using VARI, (**b**) area of stressed and healthy oil palms

frond, which causes the fronds to bend or break. Eventually, new fronds are deformed, resulting in death of young oil palms. Rhinoceros beetle infection causes distortion in the shape of oil palm crowns due to damage to the fronds. Figure 7.11 shows the pattern of the oil palm crown. In the case of healthy oil palm, the crown is star-shaped. However, due to infection caused by rhinoceros beetles, the crown became distorted. Therefore, crown shape information is useful to identify healthy and diseased oil palm plants. The crown diameter of distorted trees is also one of the proxies to identify the diseased plant. Plants can be considered diseased if the tree is distorted and crown diameter is small. To confirm the healthy plants and diseased plants, we further investigated the NDVI value of oil palms in the study area. Figure 7.12 shows the NDVI variations in oil palm plants. The NDVI value is low in diseased plants.

Fig. 7.10 Ortho-image of oil palm plantation area captured using MicaSense RedEdge sensor

Fig. 7.11 Ortho-image of oil palm plantation area with crown shape and diameter

7.5 Conclusion

The scope of UAVs is rapidly expanding. Examples of the various applications of UAVs include large-scale real estate, precision agriculture (precision farming), forestry, etc. Drone technology is becoming increasingly sought-after for various purposes by various stakeholders as a low-cost/high-yield asset. UAVs can be seen as an alternative to traditional remote sensing due to advantages such as low cost, high resolution, and time-efficient methods. This study shows that the use of UAV

Fig. 7.12 Variation of NDVI in healthy and diseased oil palm plants

data can significantly contribute toward oil palm plantation management, biophysical parameter estimation, as well as planning for the development of plantation sites and estimating the costs and benefits. Results from this study clearly show that collection of multispectral imagery from a UAV provides us with useful data to monitor oil palm health with the use of VARI, change in crown shape, and NDVI as an indicator from rhinoceros beetle disease outbreak. We conclude that UAVs may have an important role to play in plantation health monitoring by providing large-scale monitoring to overcome the limitations of ground surveys with small-scale area and infrequent visits. Presently, research on integrating satellite information and UAV ortho-image is still in progress. Comparing the scale of information that can be obtained from both data sets is important for detecting the physiological stress that results from disease or pest outbreaks.

References

Ahmed A, Ibrahim A, Hussein S (2019) Detection of palm tree pests using thermal imaging: a review. In: Machine learning paradigms: theory and application. Springer, Cham, pp 253–270

Alberts K (2012) Landsat data characteristics and holdings. A presentation of USGS landsat ground system lead. Accessed 15 May 2016. http://www.slideserve.com/keahi/landsat-data-characteristics-and-holdings

Almeida DRA, Broadbent EN, Zambrano AMA, Wilkinson BE, Ferreira ME, Chazdon R et al (2019) Monitoring the structure of forest restoration plantations with a drone-lidar system. Int J Appl Earth Obs Geoinf 79:192–198

Attarzadeh R, Momeni M (2012) Object-based building extraction from high resolution satellite imagery. Int Arch Photogramm Remote Sens Spat Inf Sci XXXIX-B4:57–60

Avtar R, Sawada H, Takeuchi W, Singh G (2012) Characterization of forests and deforestation in Cambodia using ALOS/PALSAR observation. Geocarto Int 27(2):119–137

Avtar R, Kumar P, Oono A, Saraswat C, Dorji S, Hlaing Z (2017) Potential application of remote sensing in monitoring ecosystem services of forests, mangroves and urban areas. Geocarto Int 32(8):874–885

Ballesteros R, Ortega JF, Hernandez D, Moreno MA (2014) Applications of georeferenced high-resolution images obtained with unmanned aerial vehicles. Part I: description of image acquisition and processing. Precis Agric 15:579–592. https://doi.org/10.1007/s11119-014-9355-8

Banu TP, Borlea GF, Banu C (2016) The use of drones in forestry. J Environ Sci Eng 5:557–562

Basiron Y (2007) Palm oil production through sustainable plantations. Eur J Lipid Sci Technol 109 (4):289–295

Bock CH, Poole GH, Parker PE, Gottwald TR (2010) Plant disease severity estimated visually, by digital photography and image analysis, and by hyperspectral imaging. Crit Rev Plant Sci 29:59–107

Breckenridge RP, Dakins M, Bunting S, Harbour JL, Lee RD (2012) Using unmanned helicopters to assess vegetation cover in sagebrush steppe ecosystems. Rangel Ecol Manag 65(4):362–370

Chong KL, Kanniah KD, Pohl C, Tan KP (2017) A review of remote sensing applications for oil palm studies. Geo-spatial Inf Sci 20(2):184–200

Fawcett D, Azlan B, Hill TC, Kho LK, Bennie J, Anderson K (2019) Unmanned aerial vehicle (UAV) derived structure-from-motion photogrammetry point clouds for oil palm (Elaeis guineensis) canopy segmentation and height estimation. Int J Remote Sens:1–23. https://doi.org/10.1080/01431161.2019.1591651

Gevaert CM, Suomalainen J, Tang J, Kooistra L (2015) Generation of spectral-temporal response surfaces by combining multispectral satellite and hyperspectral UAV imagery for precision agriculture applications. IEEE J Sel Top Appl Earth Obs Remote Sens 8:3140–3146. https://doi.org/10.1109/JSTARS.2015.2406339

Guerra-Hernández J, Gonzalez-Ferreiro E, Sarmento A, Silva J, Nunes A, Correia AC et al (2016) Using high resolution UAV imagery to estimate tree variables in Pinus pinea plantation in Portugal. For Syst 25(2):eSC09

Henson IE, Harun MH, Chang K (2008) Some observations on the effects of high water tables and flooding on oil palm, and a preliminary model of oil palm water balance and use in the presence of a high water table. Oil Palm Bull 56:14–22

Hojas-Gascón L, Belward A, Eva H, Ceccherini G, Hagolle O, Garcia J, Cerutti P (2015) Potential improvement for Forest cover and Forest degradation mapping with the forthcoming Sentinel-2 program. Int Arch Photogramm Remote Sens Spat Inf Sci 40(7):417–423

Hou J, Li L, He J (2016) Detection of grapevine leafroll disease based on 11-index imagery and ant colony clustering algorithm. Precis Agric 17(4):488–505

Inoue T, Nagai S, Yamashita S, Fadaei H, Ishii R, Okabe K, Taki H, Honda Y, Kajiwara K, Suzuki R (2014) Unmanned aerial survey of fallen trees in a deciduous broadleaved forest in eastern Japan. PLoS One 9(10):e109881

Jaud M, Passot S, Le Bivic R, Delacourt C, Grandjean P, Le Dantec N (2016) Assessing the accuracy of high resolution digital surface models computed by PhotoScan® and MicMac® in sub-optimal survey conditions. Remote Sens 8:465. https://doi.org/10.3390/rs8060465

Kalantar B, Bin Mansor S, Sameen MI, Pradhan B, Shafri HZM (2017) Drone-based land-cover mapping using a fuzzy unordered rule induction algorithm integrated into object-based image analysis. Int J Remote Sens 38(8–10):2535–2556

Li L, Zhang Q, Huang DF (2014) A review of imaging techniques for plant phenotyping. Sensors 14:20078–20111. https://doi.org/10.3390/s141120078

Li W, Dong R, Fu H (2019) Large-scale oil palm tree detection from high-resolution satellite images using two-stage convolutional neural networks. Remote Sens 11(1):11

Liaghat S, Ehsani R, Mansor SA, Shafri HZM, Meon S, Sankaran S, Azam SHMN (2014) Early detection of basal stem rot disease (Ganoderma) in oil palms based on hyperspectral reflectance data using pattern recognition algorithms. Int J Remote Sens 35(10):3427–3439. https://doi.org/10.1080/01431161.2014.903353

Liang YH (2008) Contributions of an agriculture engineer in the oil palm plantation. JURUTERA, 11–13. http://dspace.unimap.edu.my/dspace/bitstream/123456789/13859/1/Feature-Oil%20Palm%20Plantation%204pp.pdf

Liau SS, Ahmad A (1993) The control of Oryctes rhinoceros by clean clearing and its effect on early yield in palm-to-palm replants. In: Proceedings of the PORIM international palm oil conference: progress, prospects challenges towards the 21st century, September 9–14, 1991, Kuala Lumpur, Malaysia, pp 396–403
Luna I, Lobo A (2016) Mapping crop planting quality in sugarcane from UAV imagery: a pilot study in Nicaragua. Remote Sens 8:500. https://doi.org/10.3390/rs8060500
Martínez J, Egea G, Agüera J, Pérez-Ruiz M (2017) A cost-effective canopy temperature measurement system for precision agriculture: a case study on sugar beet. Precis Agric 18(1):95–110
Mohamad R, Mohayidin MG, Wibaya W, Juraimi AS, Lassim MM (2010) Management of mixed weeds in young oil-palm plantation with selected broad-spectrum herbicides. Pertanika J Trop Agric Sci 33:193–203
Mostafa SA, Hazeem AA, Khaleefahand SH, Mustapha A, Darman R (2018) A collaborative multi-agent system for oil palm pests and diseases global situation awareness. In: Proceedings of the future technologies conference. Springer, Cham, pp 763–775
Navarro J, Algeet N, Fernández-Landa A, Esteban J, Rodríguez-Noriega P, Guillén-Climent M (2019) Integration of uav, sentinel-1, and sentinel-2 data for mangrove plantation aboveground biomass monitoring in Senegal. Remote Sens 11(1):77
Potineni K, Saravanan L (2013) Natural enemies of oil palm defoliators and their impact on pest population. Pest Manag Hortic Ecosyst 19:179–184
Pullanagari R, Kereszturi G, Yule I (2016) Mapping of macro and micro nutrients of mixed pastures using airborne AisaFENIX hyperspectral imagery. ISPRS J Photogramm Remote Sens 117:1–10
Sandino J, Pegg G, Gonzalez F, Smith G (2018) Aerial mapping of forests affected by pathogens using UAVs, hyperspectral sensors, and artificial intelligence. Sensors 18:944
Severtson D, Callow N, Flower K, Neuhaus A, Olejnik M, Nansen C (2016) Unmanned aerial vehicle canopy reflectance data detects potassium deficiency and green peach aphid susceptibility in canola. Precis Agric 17(6):659–677
Shahbazi M, Theau J, Menard P (2014) Recent applications of unmanned aerial imagery in natural resource management. GIScience Remote Sens 51(4):339–365
Shafri HZM, Hamdan N, Izzuddin Anuar M (2012) Detection of stressed oil palms from an airborne sensor using optimized spectral indices. Int J Remote Sens 33(14):4293–4311
Shamshiri RR, Hameed IA, Balasundram SK, Ahmad D, Weltzien C, Yamin M (2018) Fundamental research on unmanned aerial vehicles to support precision agriculture in oil palm plantations. In: *Agricultural robots-fundamentals and applications*. IntechOpen, London
Shang K, Zhang X, Sun Y, Zhang L, Wang S, Zhuang Z (2015) Sophisticated vegetation classification based on feature band set using hyperspectral image. Guang pu xue yu guang pu fen xi= Guang pu 35(6):1669–1676
Shao GF (2012) Remote sensing: encyclopedia of environmetrics. Wiley, Chichester, pp 2187–2193
Song Y, Sun H, Li M, Zhang Q (2015) Technology application of smart spray in agriculture: a review. Intell Autom Soft Comput 21(3):319–333
Stek TD (2016) Drones over Mediterranean landscapes. The potential of small UAV's (drones) for site detection and heritage management in archaeological survey projects: a case study from Le Pianelle in the Tappino Valley, Molise (Italy). J Cult Herit 22:1066–1071
Tattaris M, Reynolds MP, Chapman SC (2016) A direct comparison of remote sensing approaches for high-throughput phenotyping in plant breeding. Front Plant Sci 7:1131. https://doi.org/10.3389/fpls.2016.01131
Wang Y, Zhu X, Wu B (2019) Automatic detection of individual oil palm trees from UAV images using HOG features and an SVM classifier. Int J Remote Sens 40(19):7356–7370

Chapter 8
Unmanned Aerial Vehicle System (UAVS) Applications in Forestry and Plantation Operations: Experiences in Sabah and Sarawak, Malaysian Borneo

Stanley Anak Suab and Ram Avtar

Abstract Unmanned aerial vehicle system (UAVS) or drone gains significant role for acquiring geospatial data especially in forestry and plantation operations. This was made possible by current advancement of consumer drone and availability of open-source UAVS technology for custom-made UAVS. The role of UAVS technology is seen as bridging the gap between field data collection and airborne and spaceborne remote sensing. UAVS offers great details of geospatial data at very high resolution with flexibility of deployment and cloud-free aerial photos. In Malaysia, the usage of UAVS is regulated by the Civil Aviation Authority of Malaysia (CAAM). The benefits of UAVS applications in forestry and plantation operations greatly improve efficiency through fast and timely geospatial data acquisition for various operational applications. In order to utilise UAVS technology, the forest and plantation owners' choices are either hiring third-party specialist or purchasing UAVS hardware and software with necessary training for own staff. UAVS workflow generally consists of flight mission planning, data acquisition, data processing and output integration into geographic information systems (GIS). In this chapter, various UAVS applications in forestry and plantation operations based on experiences in Sabah and Sarawak are discussed. UAVS applications are infrastructure management, roads, nursery, boundary and encroachment monitoring, nursery management and high conservation value areas (HCV).

Keywords UAV · Forestry · Plantations · Operations

S. A. Suab (✉) · R. Avtar
Faculty of Environmental Earth Science, Hokkaido University, Sapporo, Hokkaido, Japan
e-mail: stan@eis.hokudai.ac.jp; ram@ees.hokudai.ac.jp

R. Avtar, T. Watanabe (eds.), *Unmanned Aerial Vehicle: Applications in Agriculture and Environment*, https://doi.org/10.1007/978-3-030-27157-2_8

8.1 Introduction

Sustainable management of forest and plantation always requires up-to-date geospatial data for their operations. Previously geospatial data were mainly acquired from remotely sensed satellite data, old topographic maps, field notes, global positioning system (GPS) data, reports and previous records, etc. The main limitation of existing data sources is the updated information. It is quite difficult to update existing database because of accessibility of forest (ground-based data collection) and cloudy condition (satellite-based data acquisition) in Borneo. The recurring cost for annual ordering of new high-resolution satellite data adds up to the forest/plantation company's overhead cost. In addition, the fieldworks for collection of field notes and GPS data are normally time-consuming and labour intensive (Avtar et al. 2013) to completely cover large forest areas such as forest timber license area or plantation blocks.

In Sabah and Sarawak, there are two types of UAV system (custom-made fixed wing and DJI Phantom series multirotor) widely used for forest and plantation management. The reasons are because both can be acquired at reasonable cost and have been proven to be very robust for use under harsh tropical rainforest conditions. Also, the running and maintenance cost of both UAVS is reasonable, compared to other commercial UAVS available in the market. The customised fixed wing UAV is hobby of locals that enables autonomous flights by fitting it with flight controller, data telemetry communication, GPS and cameras for recording aerial photos and videos. An earlier version of a reliable open-source flight controller hardware was the ArduPilot Mega (APM)/HKPilot Mega introduced in 2012 (discontinued in 2017). Open-source firmware and Ground Control software, namely, ArduPilot and Mission Planner, were also made available by the ArduPilot Development Team and Community through (www.ardupilot.org) (ArduPilot DevTeam 2019). The more advanced successor named Pixhawk was introduced in the following year. This open-source flight controller and its software are keeping pace with the updates and its capabilities with the help of a growing number of UAV developers contributing to the expanding advancement.

Meanwhile the multirotor DJI Phantom 1 was first introduced in 2013 and become one of the best airborne remote sensing platforms for aerial surveys. Several key reasons why the DJI Phantom is very successful are its comparatively reasonable pricing (starting from USD1000), practicality, advanced features, continuous technology updates, proven stability and reliability. The 2018 Drone Market Sector Report done by Skylogic Research shows that DJI has an estimated 74% of the market share this year and 72% in 2017 with DJI Phantom 4 series taking 29% of the overall share of DJI brand (www.uavcoach.com). This chapter describes the examples of UAV applications in forestry and plantation operations based on the authors' previous work experiences with UAVs since year 2014.

8.1.1 Unmanned Aerial Vehicle System (UAVS) Components

The unmanned aerial vehicles (UAVs) commonly known to the public as drones are the main component of the UAV system. In addition, two other components of the UAV system include well-trained crew operators and the Ground Command Control hardware and software. All three components are very important to ensure a successful operational and application of the UAVS technology into the existing workflow of natural forest and plantations (forest, rubber, oil palm, etc.). Figure 8.1 shows various components of unmanned aerial vehicle system. Much recent UAV type is the vertical take-off and landing (VTOL) platform which is still under testing to see the long-term suitability for use in the tropical forest conditions.

8.1.2 Unmanned Aerial Vehicle System (UAVS) Regulations and Safety in Malaysia

It is important for the UAVS operators to know and follow the regulations by the Civil Aviation Authority of Malaysia (CAAM) (formerly known as Department of

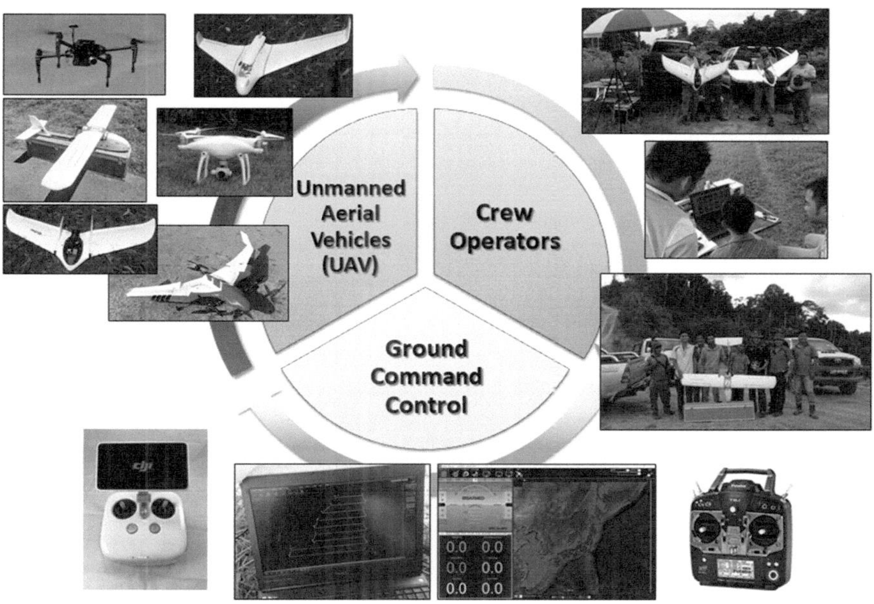

Fig. 8.1 Components of an unmanned aerial vehicle system

Civil Aviation, DCA) for their safety in conducting aerial survey works. The CAAM has made an effort to regulate the operations of UAVS in order to integrate it into the national airspace. In 2008 the CAAM has issued an Aeronautical Information Circular (AIC) titled "Unmanned Aerial Vehicle (UAV) Operations in Malaysian Airspace". According to CAAM, UAV is defined as an aircraft and its associated elements which are operated with no pilot onboard.

There are three categories of UAV under the CAAM regulations. The first one is a small unmanned aircraft system which weighs not more than 20 kg and does not require registration or obtain a certificate of airworthiness. The second category is UAV used for data acquisition or surveillance. The third category includes unmanned aircraft systems of more than 20 kg which require registration and need to obtain a certificate of airworthiness. There is no authorisation needed if you intend to fly a small UAV below 400 feet altitude. However, for the purposes of aerial survey works (means an aircraft operation in which an aircraft is used to provide specialised services in agriculture, construction, photography, surveying, observation and patrol, search and rescue, aerial advertisement and other similar activities), permission is required and fees to be paid to CAAM.

Meanwhile for safety purposes CAAM has set rules for no-fly zones in airspace classes A, B, C and G and within an aerodrome traffic zone (Table 1). In addition, no-fly zones are already included in the standard settings of all UAV Ground Control software. Safety aspects are normally included as part of the training provided by the vendor when acquiring UAV units. The scope of safety includes the operating hazards to the operators and the general public. However, it is emphasised that the operators should always be responsible for safety measures and abide by the established rules when operating the UAV. Table 8.1 shows the classification of airspace by Air Traffic Services (ATS) Malaysia.

Table 8.1 Air traffic services airspace classification in Malaysia

Classification	Levels	Airspace
A	FL460 FL 250	FIR (including ATS routes)
A	FL 250 FL 150	ATS routes and TMAs
B	FL 150 10, 000 feet ALT	
C	10,000 feet ALT Lower limit	
C	Upper limit Ground/sea	CTRs and ATZs
G	Below FL 250	Uncontrolled airspace

Source: Civil Aviation Authority of Malaysia (CAAM)

Abbreviations: *ALT* altitude, *ATS* Air Traffic Services, *ATZ* aerodrome traffic zone, *CTR* controlled traffic region, *FIR* flight information region, *FL* flight level, *TMA* terminal control area

8.1.3 The UAV Types

Two major types of UAVs used in this application are custom-made fixed wing and ready-to-fly (RTF) DJI multirotors especially the Phantom 4 series. The custom-made UAV components were DIY made using market available hobby remote control (RC) aircraft airframe mainly with delta wing (three channels and tailless) and normal aircraft like aircraft wing (four channels with tail). The RC aircrafts were assmbled and fitted with the open-source flight controller set and were set tuned as UAV, while the DJI Phantom multirotors were acquired from local authorised distributors of the DJI products.

The open-source flight controller set consists of the Flight Controller Autopilot (Pixhawk), telemetry for data transmission with ground control, power module and GPS module. Meanwhile the firmware installed in the flight controller and the ground control Mission Planner software is also open source. The customised fixed wing UAV can use RGB cameras such as Canon S100, S110, Sony RX100 series, Sony Nex5R series, Sony A5000/A6000 series and thermal and multispectral MicaSense RedEdge. Table 2 shows details of fixed wing and the multirotor UAVs. This table can be used to decide suitability of UAVs for various applications.

8.1.4 Pros and Cons of UAVS Applications

Unmanned aerial vehicle system (UAVS) is an advanced version of remote sensing tool for improving accuracy and efficiency by providing fast, detailed and timely geospatial data. The system allows a synoptic view with high-resolution data acquisition of present ground scenario and what happened after implementation of operational practices in natural forestry and plantation operations. UAVs are versatile working tools which can be deployed anytime anywhere within the working area. Using UAV as the "workhorse" for gathering spatial data also provides an opportunity for gathering sets of temporal data which result in rich spatial information as inputs for the existing GIS database. Figure 8.2 shows use of various spatial data collection platforms and their resolution and area coverage. Satellite is useful to cover huge area with coarse resolution, and UAV is useful to cover small area with fine resolution.

Applications of new technology into existing workflow always require the staff to acquire new skills and knowledge through learning by doing and formal training. The challenges are to develop qualified manpower who are able to handle data collection tasks using UAV in the field. This will require some amount of time and also strong commitment of the staff to master the art of using UAVs and to integrate their outputs into the existing workflows.

Table 8.2 Details of the UAV used

Properties	Fixed wing UAV (custom-made)	Multirotor
Type of platform	Flying wing and	DJI Phantom 4 series (ready to fly)
	Fixed wing frame	
Operation advantages	Long range and more flight hours (up to 50 minutes per flight)	Vertical take-off and landing
	Covers more area (100–300 hectares/flight)	Excellent platform for capturing videos
	Freedom of payload options (RGB cameras or multispectral)	Great for very low-altitude high-res. detailed data acquisitions (e.g. 360 degrees of a building inspections)
	Customisable	
Operation disadvantages	Needs space for take-off and landing	Short flight hours (20 minutes), short range and covers small area
		Fixed camera payload
	Need specialised knowledge and skills to do maintenance, repairs and troubleshooting	Adding NDVI or thermal cameras will add up to the weight and shorter flight time
		Maintenance, repairs and troubleshooting have to be done by authorised DJI technicians
Hardware	Remote control hobby aircraft components with electric propulsion system	Ready to fly (consumer drone grade) with GPS
		Lightbridge technology
	Fitted with open-source flight controller set (Pixhawk) with GPS, telemetry modem	12.4 meegapixels for P4 and
	Custom payload 12, 16 or 20 megapixel cameras	20 megapixels for P4Pro
		Obstacle avoidance
Software	Open-source firmware	DJI firmware
	Open-source Mission Planner	Litchi
		Pix4DCapture, Drone Deploy & DJI GO
		Other third-party apps
Usage	Mapping larger area of interest (AOI)	Mapping smaller area of interest (AOI)

8.1.5 UAVS: A Cost-Effective Working Tool

There are two options for UAV technology applications for collection of geospatial data which are (a) hiring a third party to carry out the task of data collection, processing and analysis and (b) purchasing UAV units and training of own staff. Basically, both options are available depending on the return of investment (ROI) of the company or users.

The first option of hiring a third party (specialist, UAV service provider) is typically the better choice for smallholder of forest plantations, rubber plantation and oil palm plantations and also for investors who also only need a "one-off" type of geospatial information for a specific area or topic of interest. This is because the

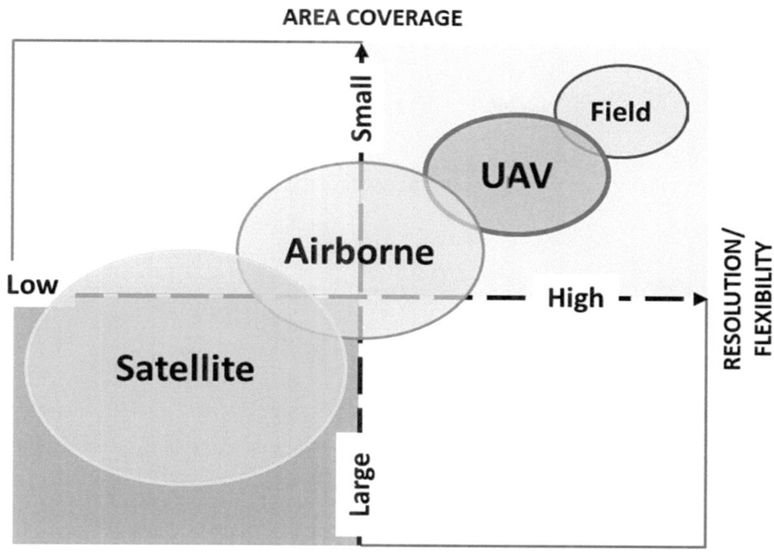

Fig. 8.2 Comparison of UAV data collection with other spatial data collection methods. (Source: Adapted from Terra Solid Point Cloud Intelligence)

specialist does not only carry out data collection and processing but mostly also can provide analytical GIS services to produce the full range of outputs needed by the client. Based on the experiences, services for field data collections and processing of indicative price ranges start from USD8.36 per hectare with a minimum area of 2500 hectares to USD3.82 for larger area of interest (AOI). In addition, the orthomosaic output interpretations and GIS analysis indicative costs are around USD3.58 to USD4.78 per hectares. Normally larger AOI coverage will allow much lower pricing which can be negotiated between the client and the service provider. Moreover, hiring a specialist is much more convenient because they have the expertise and experience to do the job right and to deliver quality information on time.

The second option of purchasing UAV units with training is also a good option for companies/users with needs for continuous data collections on a daily or weekly basis. In this case, the GIS staff needs to learn and adopt the UAV for use as an additional source of data and information. Normally the company/user chooses to have both fixed wing and multirotor UAV at the same time because one UAV can complement the other UAV function during fieldworks. Besides the UAV operations training, the staff also must be trained for basic troubleshooting and maintenance of the UAV units. Based on authors' previous experience, the cost for acquiring both fixed wing and multirotor UAV including training will range from around USD7,800 to USD13,000 depending on how many UAV units including type of sensor, accessories and spare parts required by the user. The user also needs to acquire the aerial photo processing software (USD3,500 for Agisoft PhotoScan Professional/Metashape and USD4,482 for Pix4D) and is required to have a good workstation computer to be used mainly for aerial photo processing (cost around USD5,000 or higher). Figure 8.3 shows the fixed

Fig. 8.3 UAV operators' training

wing UAV and multirotor UAV DJI Phantom 4 Pro with the participants during one of the training courses in Malaysia for oil palm applications.

8.2 UAVS Operations Workflow

The UAV works start with a designated task issued by the management. It is usually based on the current priorities with certain type of information required for decision-making. After the objectives have been determined, the manager in charge of the geospatial data will identify the exact location and extent of the area of interest (AOI). The AOI is issued in the format of GIS shapefile or KML which directly can be used by UAV operator to plan for the data collection fieldworks. Figure 8.4 shows the outline of the UAV workflow.

8.2.1 Flight Mission Planning

At this stage the UAV operator will make a plan for the flight path, using the Mission Planner software with the AOI shapefiles. Details such as flight height, side and front overlapping, required ground sampling distance (GSD) resolution and camera

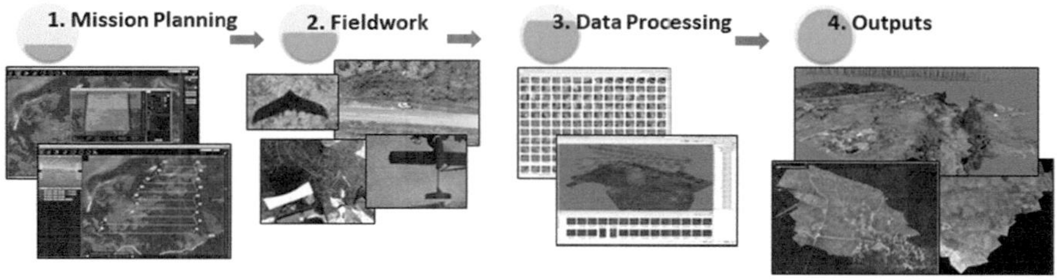

Fig. 8.4 UAV workflow

calibrations have to be taken into consideration when planning a flight mission. However, ground sampling distance (GSD) in the Mission Planner software is automatically calculated based on the selected flight parameters. Also, it is crucial for the operators to have good knowledge about the field conditions (e.g. topography, especially for selection of the best possible spot for take-off and landing). Based on the flight planning, the operator should be able to estimate the total number of flight missions and days required to complete the data collection including backup day in the case of bad weather conditions. Subsequently, the operator needs to plan for logistic matters such as other equipment needed, spare parts and extra UAV units as a backup, depending on the duration of the missions and accessibility of the site. The success of the data collection mission depends much on good flight mission planning.

8.2.2 *Data Acquisition*

The next step is field deployment of staffs and aerial survey equipment. Normally a team of three staff with responsibility as UAV pilot, mission controller and spotter/observer are deployed for the fieldwork. The best practice before conducting the aerial survey is to determine the ground control points (GCP) within the AOI. This is especially needed if the area does not have reliable and accurate GIS base data like compartment/block boundary, road network, reference point information or other reliable geometrically corrected reference data for quality control. UAV safety checks are done before, during and after data acquisition. During the data acquisition process, the team needs to be able to use their best judgement and decision of possibility to abort the flight mission due to sudden bad weather or abnormal behaviour of the UAV unit. This is to avoid any unwanted incident to happen. The best time based on experiences for aerial survey in the tropics is normally between 9 am and 3 pm to ensure optimal illumination, to minimise shadow effect and to avoid high winds. An activity log is normally recorded for each fieldwork, together with a flight check list and comments form to be filled in. This is a useful record for equipment maintenance purposes.

8.2.3 Data Processing

The data processing works are mostly carried out using Agisoft PhotoScan Professional software or Pix4D. Generally, the steps in the data processing include the following: load aerial photos, manually determine quality and remove photos of low quality, align photos, import ground control points (GCPs) if necessary and build a dense point cloud, mesh, texture, DEM from dense cloud and lastly the orthomosaic. Outputs of the processed aerial photos are orthomosaic (GeoTIFF and KMZ Google Earth format), digital elevation model (DEM), point clouds and generated processing report.

8.2.4 Output Integration in Geographic Information System (GIS) and Google Earth Viewing

The orthomosaic aerial photos, digital surface model (DSM), point clouds and digital elevation model (DEM) are imported in LAS point elevation, GeoTIFF for GIS and in Google KMZ format for easy and quick viewing purposes. The KMZ format enables possible data sharing by sending them through email attachment because of their small size. The output quality is assessed prior to integration into the existing GIS dataset for analysis purposes. During the processing of the aerial photo, geometric corrections are normally applied using existing GIS layers or using field ground control points (GCPs) obtained from the site. This is important to ensure an utmost accuracy of the geospatial data. Figure 8.5 illustrates an example of post-

Fig. 8.5 Example of post-processing accuracy assessment of orthomosaic aerial photo

processing accuracy assessment of orthomosaic aerial photo using ground control points (GCPs) collected from real-time kinetics (RTK).

8.3 Case Studies

8.3.1 UAVS Applications in Forestry and Plantation Infrastructure Management

Infrastructures in natural forestry and plantation operations typically include roads, bridges, office, entrance gates, camps, loading/unloading area, weighbridge facility, log yard, workshop, nursery, quarries, gravel stockpile, fire towers and many additional site feature landmarks. In addition, availability of UAVS technology has made it possible to carry out regular monitoring and ad hoc inspections of work progress on-site. This includes identifying potential work hazards and risk of work-related accidents for development of safety precaution measures, e.g. during infrastructure development and maintenance work involving heavy machineries. The work progress and condition of infrastructure development also can be easily inspected visually from the processed aerial photo or aerial video to acquire detail without the need to have physical contact with the objects on-site. Asset inventories and cross-checks of asset status, location and distribution can be easily done in a fast and efficient manner. Figure 8.6 shows various examples of infrastructure management in plantation and forest sites.

8.3.2 UAVS Applications in Forestry and Plantation Road Works

Road constructions are prerequisite to forestry and plantation operations. A well-designed, constructed and maintained roads are important to reduce and prevent severe impacts of forestry and plantation operations on the environment sustainability. Also important on the operation itself is costing because road construction and maintenance cost in the tropical is often higher (www.ifmconsult.com).

In Sabah and Sarawak, the roads, forest roads, bridges and culverts which are established for the forest operations or plantations are also benefiting the local communities. In Sabah sustainable forest management (SFM) licensing agreement has conditions that include local communities. Meanwhile in Sarawak it was under the Sarawak Land Code (Cap 81) (Revised Law of Sarawak 1958) which is the main legislation determining land tenure and administration in the state that provides for native area land, native customary land and three other categories of land use (ITTO 2004).

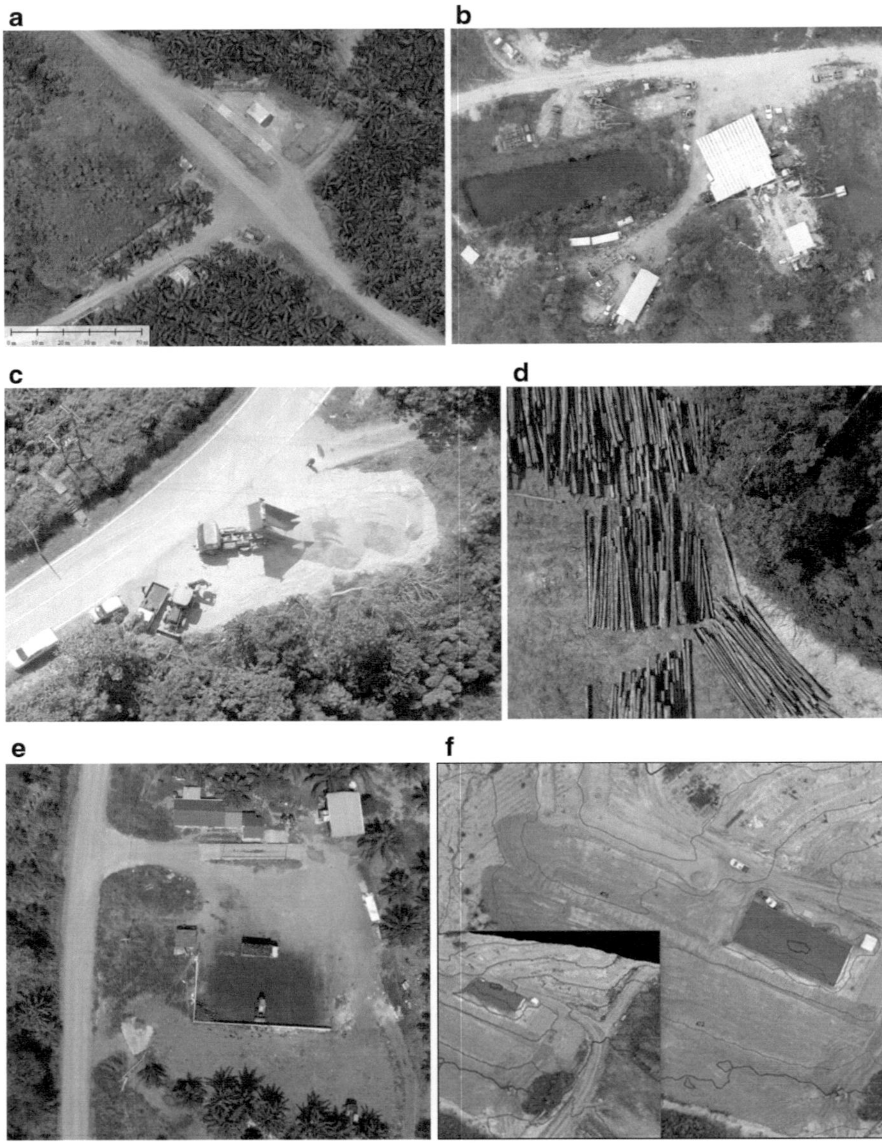

Fig. 8.6 (**a**) Weighbridge and entrance gate, (**b**) workshop, (**c**) gravel stockpile, (**d**) log yard, (**e**) loading/unloading area, (**f**) site office planning

Therefore, it is important for the Forest Management Unit (FMU) licensee and the plantations licensee to keep the roads safe for use and maintained. The previous practice for road planning uses topographic maps or digital terrain model and designs the roads using GIS. Meanwhile for road maintenance, inspection was

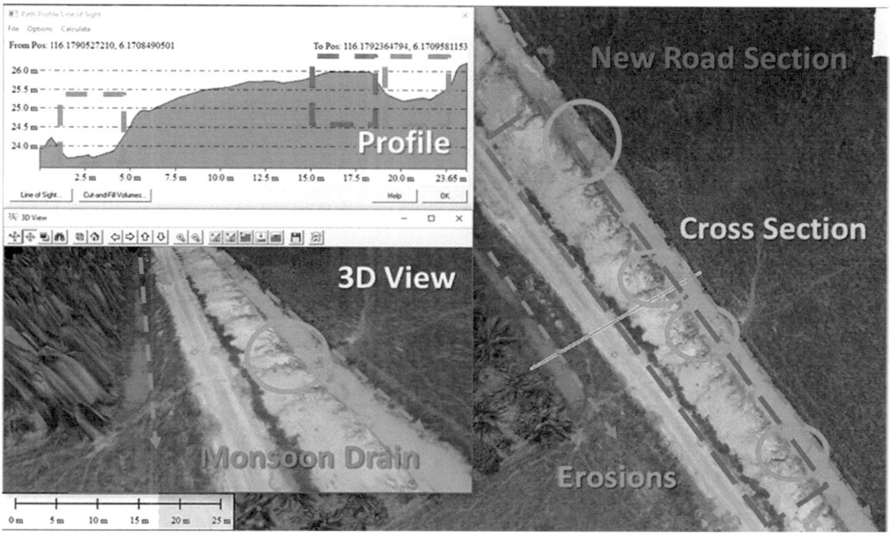

Fig. 8.7 Road work inspection in two- and three-dimensional view

done visually on-site. Availability of UAVS technology helped shortcut the obviously time-consuming and costly manual works and also leverage the capability of GIS for road planning by providing information-rich geospatial details from orthomosaic aerial photo.

UAVS application is a very effective solution for surveying, monitoring and recording the current state of the road works. In one of the cases, there was inspection on road conditions of newly added road section, monsoon drain, erosion and effects of overtime one-sided compaction of the old road section. Details of the road conditions can be seen in Fig. 8.7 in two- and three-dimensional perspectives.

An up-to-date high-resolution aerial orthomosaic data enables viewing advantage in two- and three-dimensional angle in great details. Such great details are definitely not visible from medium-resolution conventional aerial photo from manned aircraft and also from high-resolution satellite imagery. Also, UAVS enables data collection coverage of bigger area compared to field survey. It is timely flexible and multi-temporal depending on the needs for accurate field information.

Another good example of application of UAVS was the case of inspection of a 12-kilometre plantation and forest road for the purpose of upgrading and maintenance works (Fig. 8.8). In this case some of the road sections are on the flood-prone low-lying topography which has already been identified using digital terrain model in GIS. Aerial photos of the area were collected in a half-day working time, and the orthomosaic outputs were produced for the GIS applications. Inspection results identified five sections of the road were badly damaged due to water erosions, and the major one is on maintenance work for a road diversion. Through the identifications, the safety department can take actions on beefing up the road safety by

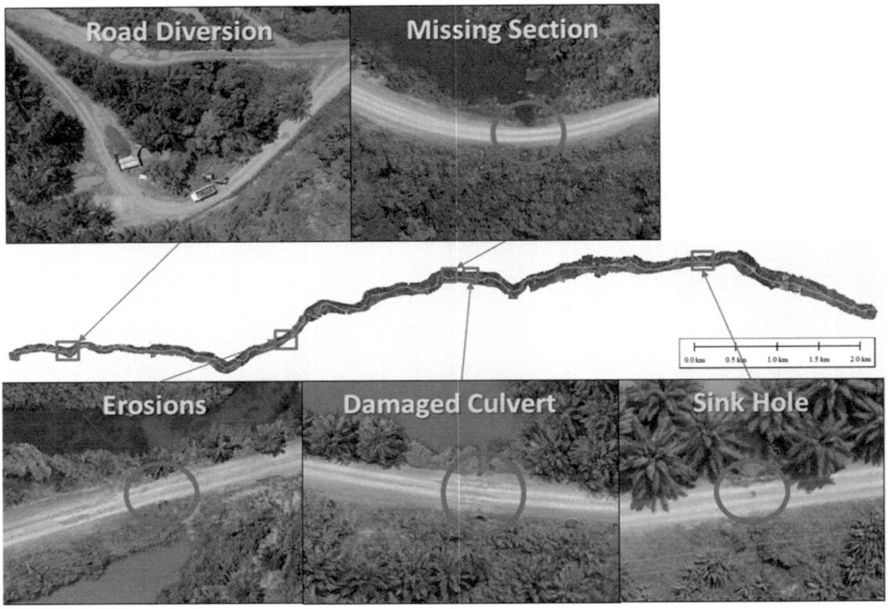

Fig. 8.8 A 12-km plantation and forest road survey

erecting temporary caution signboard to the affected area. In addition, information such as how long the stretch of road needed for repair, what kind of damages, and what kind of repair is needed can easily be quantified and identified. The nature of road damages and conditions of the scenario are also easily understood and visually described using 2D and 3D data for reporting and discussion purposes; thus costing and appropriate measures can be finalised in no time for the management.

8.3.3 UAVS for Nursery Management

Nursery is also an important part of forestry and plantation operations. Common plants grown in the nursery are oil palm, commercial fast-growing timber trees (acacia, eucalyptus, kelampayan, etc.) and rubber trees, and indigenous tree species are also grown for forest rehabilitation. UAVS data are very useful especially for nursery planning purposes. For example, the layout arrangements of seedlings are easily monitored for estimation amount of seedling stocks, progress and capacity of productions. The number of workers needed to take care nursery productions also can be estimated. Estimation of raw materials such as compost, soil, fertilisers, insecticides, nursery apparatus/equipment and water needed in the nursery operations also can be done fast. Figure 8.9a, b shows the nursery layout and seedling stocks.

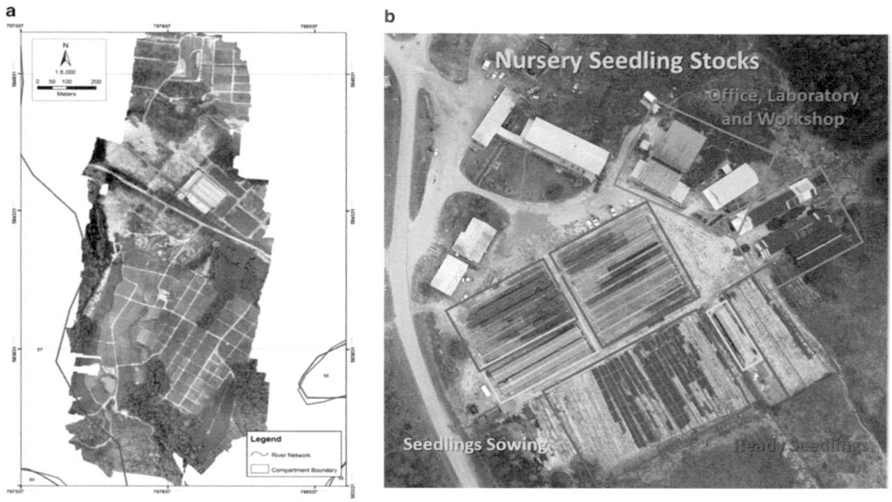

Fig. 8.9 (**a**) nursery layout, (**b**) seedling stocks

8.3.4 *UAVS for Boundary and Encroachment Monitoring*

Boundary monitoring in forest and plantation is carried out to ensure all work activities are done within the designated approved license area and also to ensure there are no encroachment activities. At the same time, boundary marking's purpose is to establish physical identification of forest and plantation compartment or blocks on the site. Boundaries are physically on ground delineation; some may have cadastral stone marks or visible markings on trees using paints, signboards, fences and clearings along the boundary buffer zones. Figure 8.10 shows plantation boundary monitoring and their encroachment.

On-site physical boundary markings provide easy guidance for navigation orientation when using hard copy maps and handheld GPS. Identifications of block/compartment ID for operational works such as harvesting, fertiliser application, pesticide applications, replanting, terracing or silviculture treatments will be carried out properly on the site without confusions. UAVS application provides useful additional information to the GIS-produced work maps, whereas orthomosaic aerial photos are used as the map background.

Similar with other UAVS applications, quantification on the encroachments can be done with details of on-site scenario which are visible in high resolution. The two examples (Fig. 8.10b, c) are clear-cutting and slash and burn encroachments inside a forest plantation. Visible is a hut constructed near the centre of the clearings with visible strips of preparation rows for some crop planting most probably rubber. The second case of slash and burn method was the fastest way of clearing which most of the time can cause wildfires especially during dry season.

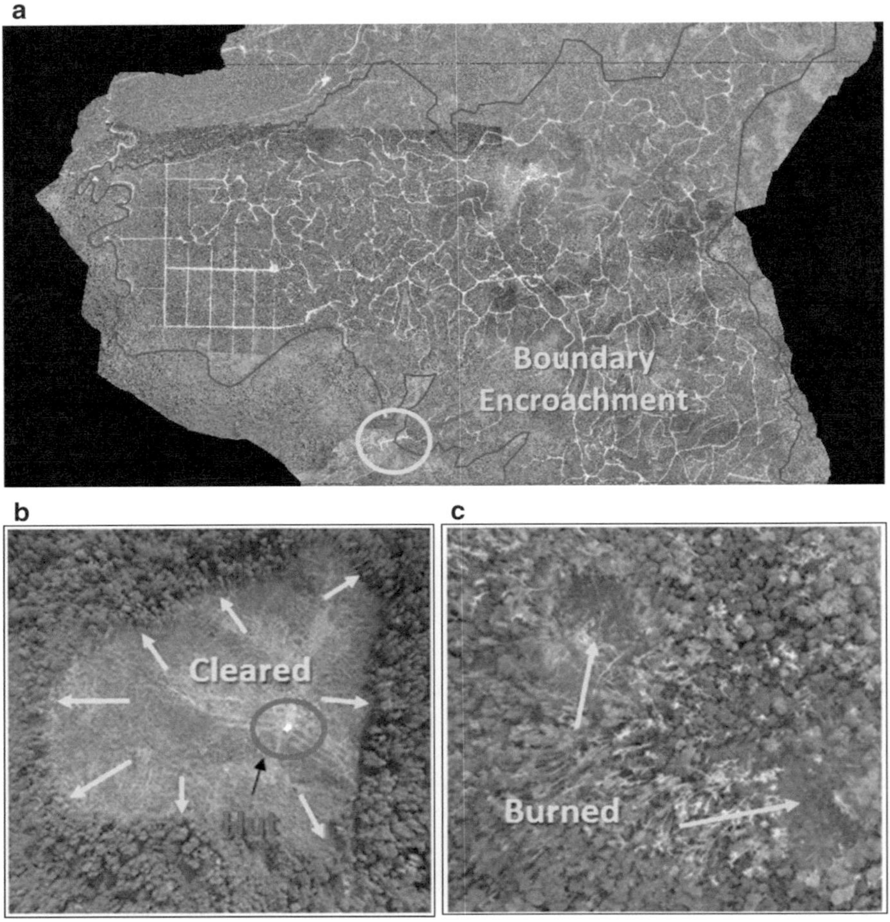

Fig. 8.10 (**a**) Plantation boundary monitoring, (**b**) clear-cutting encroachment, (**c**) slash and burn encroachment

8.3.5 *UAVS for High Conservation Value (HCV) Areas: Management and Monitoring*

High conservation value (HCV) areas are those where biological, ecological, social or cultural values are exceptionally significant or critically important within a landscape (WWF 2009). The HCV approach aims to ensure the areas are protected while allowing economic development and agricultural production (rspo.org). The concept of HCVF was initially developed by the Forest Stewardship Council (FSC) for use in the forest management certification (WWF 2009), and the previous HCVF toolkit for Malaysia was published by WWF-Malaysia in 2009.

Fig. 8.11 High conservation value buffer inside forest replanting area

A new common HCV Malaysia toolkit was developed in 2018 by the HCV Malaysia Steering Committee which was jointly membered by the Forest Stewardship Council (FSC) Malaysia, Malaysian Palm Oil Association (MPOA), Malaysian Palm Oil Certification Council (MPOCC), Malaysian Timber Certification Council (MTCC), Roundtable on Sustainable Palm Oil (RSPO), Roundtable on Sustainable Biomaterials (RSB) and WWF-Malaysia and was facilitated by Profest (www.rspo.org).

The new toolkit highlighted six high conservation values (HCVs) comprised of species diversity, landscape-level ecosystems and mosaics, ecosystems and habitats, ecosystem services, community needs and cultural values (HCV Malaysia Toolkit Steering Committee 2018). These HCV areas are excluded from the working areas and treated separately. Appropriate buffer zone areas were demarcated to isolate the HCVs, and necessary ground markings were done along the buffer perimeters. This is to ensure the working contractors will not encroach the protected HCV zones. Figure 8.11 shows the high conservation value areas identified using UAV data and their importance.

8.4 Conclusions

The current advancement and availability of both consumer and open-source unmanned aerial vehicle system (UAVS) technology opens the opportunity for various applications in the forestry and plantation operations. Acquisition of high-resolution aerial photos using UAVS helps bridge the gap between manual fieldwork data collection and existing remote sensing platform such as airborne and satellite platform. Manual fieldwork data collection is time-consuming and costly. Meanwhile synergistic use of in situ and UAV data can provide better decisions in forestry and plantation operations. In Malaysia some of the private companies are using their own UAVs to collect high-resolution data, and they can improve data collection procedure by following standard methodology. UAVS technology is still costly for some stakeholders; therefore developing standard UAV at a low cost is important to ensure everyone will be benefited from this technology. We also need to ensure sustainable practices in future application to deliver high-end results for more efficient and precise sustainable forest and plantations.

References

ArduPilot DevTeam (2019) Plane home. http://ardupilot.org/plane/. Accessed on 8 Dec 2018

Avtar R, Takeuchi W, Sawada H (2013) Full polarimetric PALSAR-based land cover monitoring in Cambodia for implementation of REDD policies. Int J Digital Earth 6(3):255–275. https://doi.org/10.1080/17538947.2011.620639

Civil Aviation Authority of Malaysia (CAAM). http://www.dca.gov.my/aviation-professionals/faq/faq-on-unmanned-aircraft-system-uas/. Accessed on 6 Nov 2018

HCV Malaysia Toolkit Steering Committee (2018) Malaysian national interpretation for the identification of high conservation values. HCV Malaysia Toolkit Steering Committee, Kuala Lumpur

ifmconsult.com (International Forest Management Consultant website). Forest Infrastructure Development. https://www.ifmconsult.com/index.php/our-services/forest-operations/forest-infrastructure-development. Accessed on 3 Jan 2019

International Tropical Timber Council (ITTC) (2004) Progress report on the study on: forest law enforcement and governance in Malaysia in the context of sustainable forest management. International Tropical Timber Organization, Interlaken, Switzerland

International Tropical Timber Organization (ITTO) Reduced Impact Logging. https://www.itto.int/sustainable_forest_management/logging/. Accessed on 3 Nov 2018

rspo.org. Roundtable on Sustainable Palm Oil news on 19 September (2018) New guidance for working with high conservation values in Malaysia. https://rspo.org/news-and-events/news/new-guidance-for-working-with-high-conservation-values-in-malaysia. Accessed on 3 Nov 2018

Sabah Forestry Department (SFD) Website on Sustainable Management –Reduced Impact Logging (RIL). http://www.forest.sabah.gov.my/discover/sustainable-management/reduced-impact-logging. Accessed on 6 Jan 2019

Terra Solid Point Cloud Intelligence, UAV LiDAR slides. https://slideplayer.com/slide/11939259/. Accessed on 10 Sept 2018

uavcoach.com, DJI Maintains 74% of market share | Three key insights from the Skylogic 2018 Drone market report. https://uavcoach.com/skylogic-2018-drone-industry-benchmark/. Accessed on 10 Sept 2018

WWF (2009) High conservation value forest (HCVF) tool Kit for Malaysia: a national guide for identifying, managing and monitoring high value conservation forests. World Wildlife Fund Malaysia, Kuala Lumpur

Chapter 9
UAV-Based Structure from Motion – Multi-View Stereo (SfM-MVS): Mapping Cliff Face of Central Crater of Mt. Miharayama, Izu Oshima, Central Japan

Toshihiro Urayama, Tatsuro Chiba, Takumi Mochizuki, Syunsuke Miura, Shino Naruke, Hisashi Sasaki, Kenichi Arai, and Hideki Nonaka

Abstract The purpose of this study is to develop a real-time measurement method for monitoring or detecting volcanic disasters using unmanned aerial vehicle (UAV). This study was supported by the Ministry of Education, Culture, Sports, Science and Technology (MEXT) Japan under "Integrated Program for Next Generation Volcano Research and Human Resource Development" project. Generally, the measurement of active volcano is dangerous because of less accessibility. But with the use of UAV human damage can be avoided compared to human exposure to danger, especially when manned aerial vehicle are used; furthermore, high-resolution imagery can be acquired quickly, which is better than satellite imagery. In general, aerial measurement is carried out by flying UAV horizontally and scanning the objects from nadir angle. However, measurement accuracy will be low since a camera cannot obtain enough data due to a big gradient caused by the crater wall. In this study, the UAV scanned inside crater from a direction approximately perpendicular to crater wall. The scanned images were used to create a 3D model of the interior of the crater. In addition, the correspondence between altitude of lava lake surface and crater volume was calculated. Therefore, it is possible to forecast lava overflow time based on the observed height of lava lake surface only.

Keywords Volcano · Crater · 3D modeling · Lava lake · Overflow prediction

---

T. Urayama (✉) · T. Chiba · T. Mochizuki · S. Miura · S. Naruke · H. Sasaki
K. Arai · H. Nonaka
Asia Air Survey, Co., Ltd., Kawasaki, Japan
e-mail: tsh.urayama@ajiko.co.jp

R. Avtar, T. Watanabe (eds.), *Unmanned Aerial Vehicle: Applications in Agriculture and Environment*, https://doi.org/10.1007/978-3-030-27157-2_9

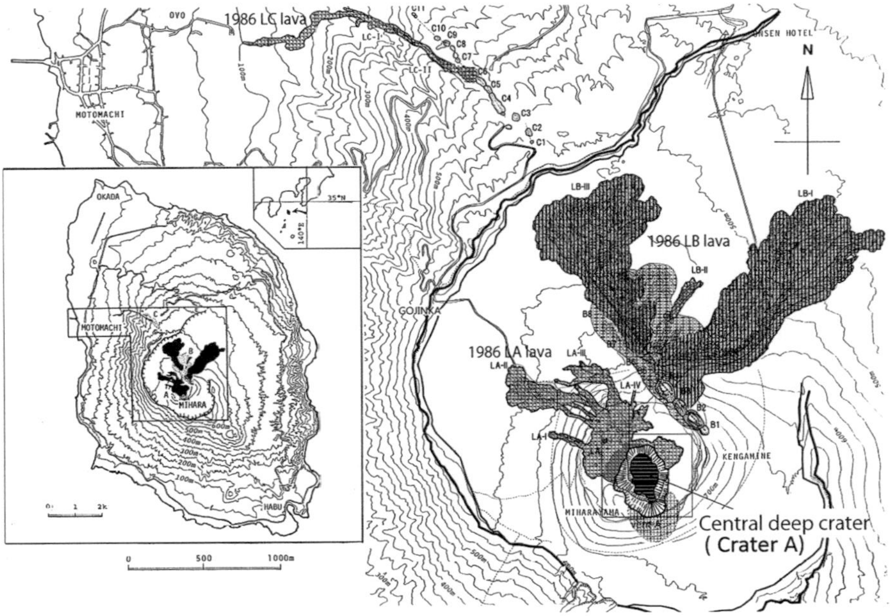

Fig. 9.1 Location of Izu Oshima Island and Mt. Miharayama

9.1 Introduction

The Ministry of Education, Culture, Sports, Science and Technology Japan (MEXT) started the "Integrated Program for Next Generation Volcano Research and Human Resource Development (http://www.kazan-pj.jp/profile)" project when Mt. Ontakesan erupted in 2014. This 10-year project will develop more effective mitigation technology for monitoring and analyzing volcanic eruptions as well as informing the public. Asia Air Survey (AAS) Co., Ltd is the first member of this project. The purpose of this study is to develop a real-time measurement method for monitoring or detecting volcanic disasters using UAV.

During a volcanic eruption, access to the vicinity of the crater is restricted. Therefore, airborne data acquisition and monitoring are required. Generally, the use of manned aerial vehicle during volcanic eruption is very dangerous. Therefore, unmanned aerial vehicles (UAVs) can be used as an alternative since they can acquire data from close range. In this study, we used an UAV for scanning the surrounding areas and interior of Mt. Miharayama central crater (Crater A, Fig. 9.1) in Izu Oshima Island. Following that, a 3D model was created based on Structure from Motion-Multi-View Stereo (SfM-MVS) technology. We calculated the correspondence between the altitude of lava lake surface and crater volume. As a result, it is now possible to forecast lava overflow time based on the observed height of lava lake surface only (Chiba et al. 2019).

Izu Oshima is an island of active volcanoes about 110 km south-southwest of Tokyo. The highest point on the island is Mt. Miharayama, with an elevation of 758 m. There is a caldera about 4 km in diameter at the summit of the mountain.

The diameter of the summit crater is about 800 m, and the diameter of Crater A is about 300 m. In the last 10,000 years, a large eruption causing hundreds million cubic meters of volcanic products at intervals of approximately 100–150 years took place. The previous large eruption occurred in 1777 and continued for 16 years, of which the volcanic products were about 0.65 billion cubic meters. The latest eruption occurred in 1950–1951 and 1986, and each volcanic product was ten millions of cubic meters (Japan Meteorological Agency). Based on this information, there are concerns that the next largest eruption would probably occur in 10 years.

9.2 Process of Izu Oshima Island 1986 Eruption

Izu Oshima 1986 eruption began on November 15th. Magma came out from the cracks generated at the edge of Mt. Miharayama central crater, and gradually landfilled the crater. The magma began to spread around the area on November 18th, then eventually flowed down the slope of the mountain, and reached the caldera floor on November 19th. This was also the case for the 1950 and 1951 eruption. Therefore, it is highly probable that the eruption would be repeated in the future (Figs. 9.2 and 9.3).

Fig. 9.2 Process of repeated eruption in the past

Fig. 9.3 Lava flow remaining at Mt. Miharayama

It is important to know the time of lava overflows for disaster prevention. In 1986, volcanologists observed in the vicinity of the crater and succeeded in predicting the lava outflows on November 19th. However, in recent years, this dangerous investigation has become impossible. Therefore, use of alternative methods will be a way forward to study volcanic disaster.

9.3 Scanning the Inside of the Crater Using UAV Camera

It is necessary for disaster prevention to know the time of lava overflows. For that, we need to know the shape inside the crater. While aeronautical laser measurements have also been carried out so far, it was difficult to measure overhangs. In topography measurement, a camera mounted on the aircraft is directed straight down. However, this method is not able to get enough information of the crater wall.

In this study, we tried photographing the crater wall by letting UAV fly over the central crater and inside the crater by manual operation. Figures 9.4 and 9.5 show the flight image, flight course and the shooting direction, respectively. The takeoff point was the crater observatory as the relative altitude is zero. We were letting the UAV fly inside the crater, tilting to adjust the camera angle to face the crater wall, and shooting rotated the UAV little by little. This operation was repeated in three steps of 0 m, −20 m and −40 m in relative altitude.

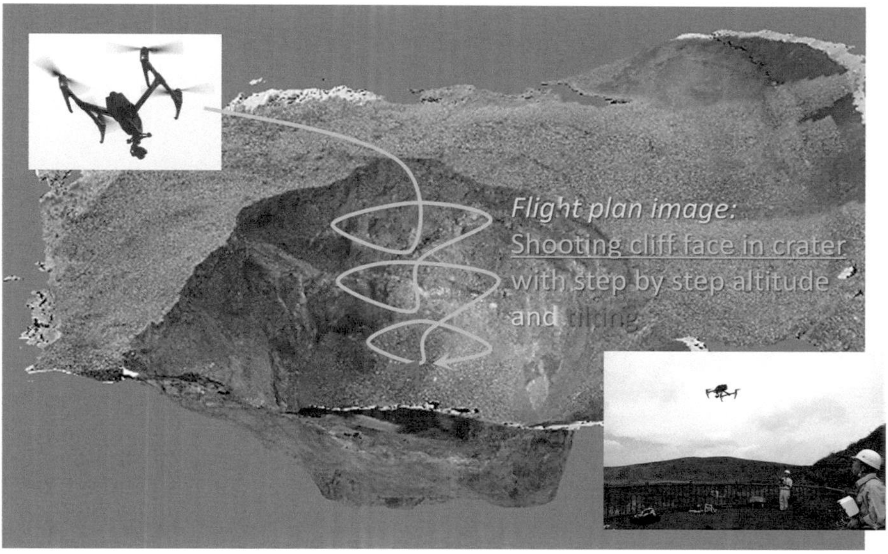

Fig. 9.4 Flight image inside the crater

Fig. 9.5 Flight course and shooting direction

The DJI Inspire 2 UAV was used for collecting data on 14th September, 2017. Zenmuse X4S (4K) camera of DJI was used to capture the pictures. The shooting time interval was 2 seconds (Fig. 9.6). A total of 400 images were acquired.

Fig. 9.6 Specification of UAV and Camera used (Background is Mt. Miharayama)

9.4 3D Modeling by UAV-Based SfM-MVS Mapping

We created a 3D model of the crater using the SfM-MVS mapping technique from the captured images (Oda et al. 2015; Fuse 2016). Agisoft Photoscan modeling software was used in this study to process the acquired images. Figure 9.7 shows the 3D model of the central crater of Mt. Miharayama.

9.5 Position Correction of the 3D Model

The 3D model was created from the position and direction of the camera and the result of image stereo matching. The position of the camera was detected from GNSS (global navigation satellite system) signals such as received GPS (global positioning system). The direction of the camera was detected from the control signal of the gimbal which was holding the camera. However, the resulting 3D model often deviates from the position and height of the actual terrain in many cases. Therefore, it was also required to correct the position of the 3D model. In this 3D model, as a result of comparison with 5 m DEM (Digital Elevation Model) of Geospatial Information Authority of Japan (GSI), there was the positional deviation of 20 m at altitude and 3 m at the horizontal position (Fig. 9.8). We corrected the positional deviation using eight GCPs (Ground Control Points). The coordinates of GCPs were obtained using Google Earth (Fig. 9.9). Figure 9.10 shows the positional deviation of the 3D model after correction. The white area of Fig. 9.10 shows the minimal positional deviation.

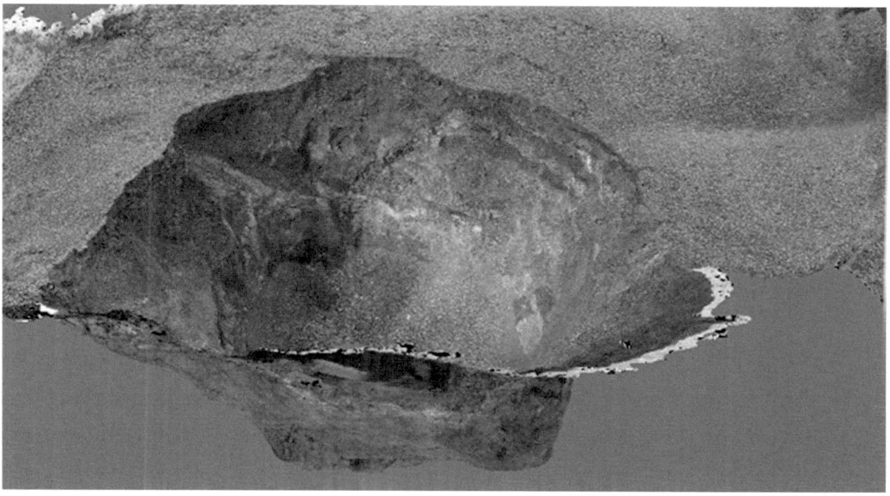

Fig. 9.7 3D model of the central crater (Crater A), Mt. Miharayama

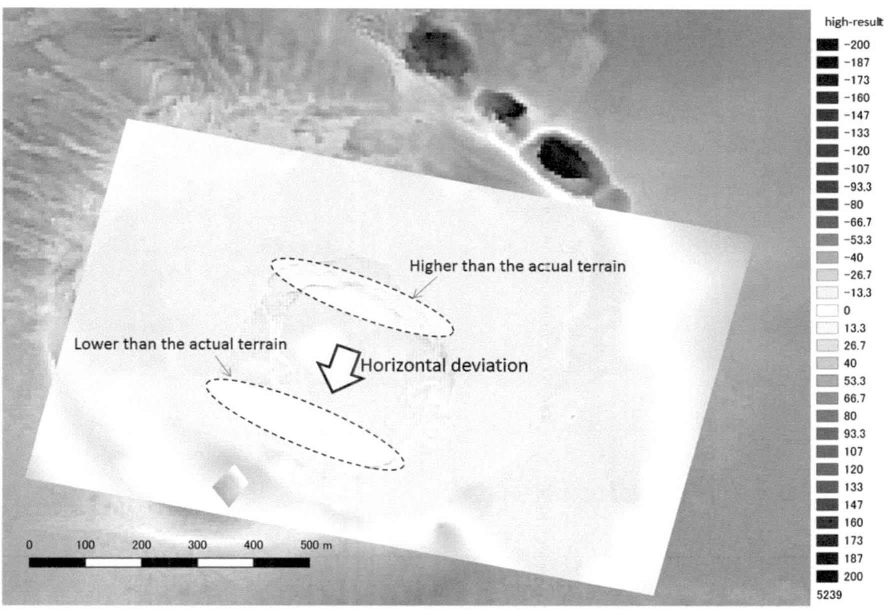

Fig. 9.8 Positional deviation of 20 m at altitude and 3 m at the horizontal position, Comparison with 5 m DEM

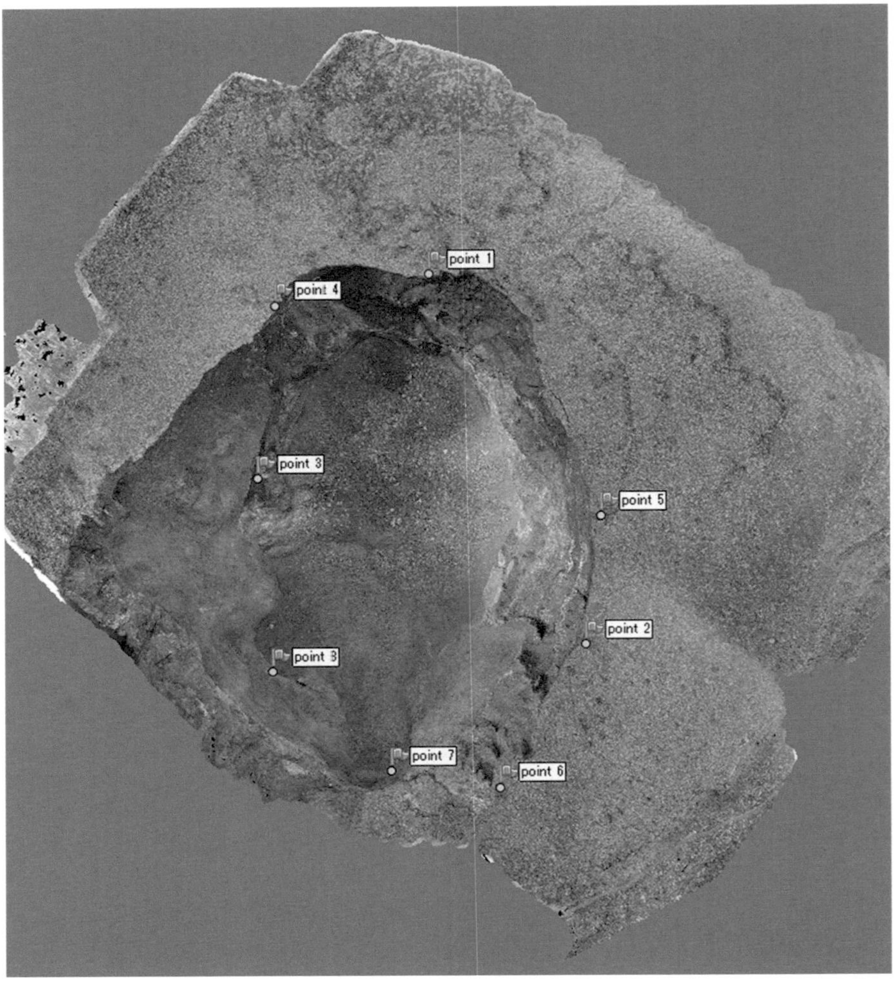

Fig. 9.9 GCPs on the 3D model of the central crater as Crater A

9.6 Measurement Result

From the position corrected 3D model of the crater, we produced a 10 cm size mesh Digital Surface Model (DSM) and the Red Relief Image Map (RRIM) (Fig. 9.11). RRIM is a novel 3D visualization technique developed by Asia Air Survey in order to represent and interpret features on the land surface, seafloor, as well as on other celestial bodies. RRIM uses multiple viewing angle geometry and a red graduated color scheme to represent and visualize terrain features more clearly.

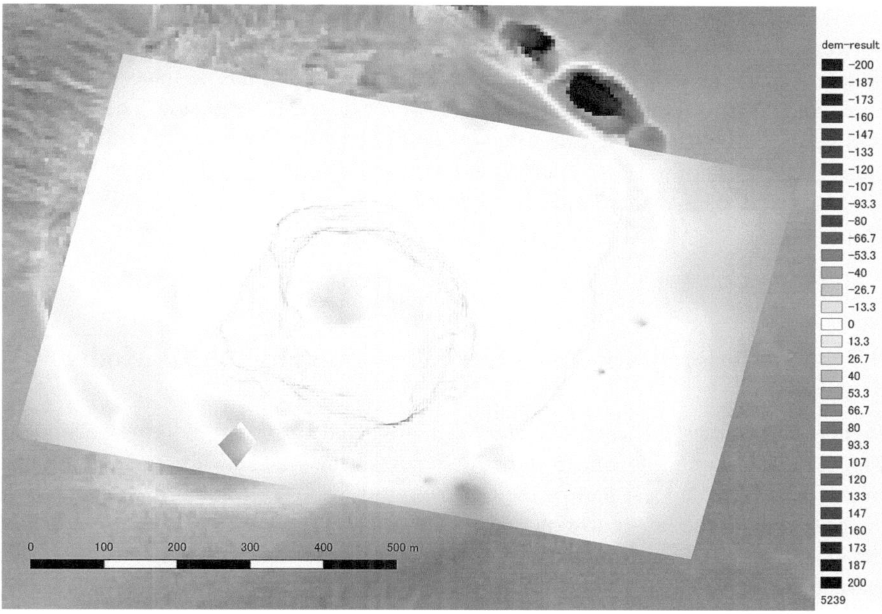

Fig. 9.10 Positional deviation of the 3D model after corrected

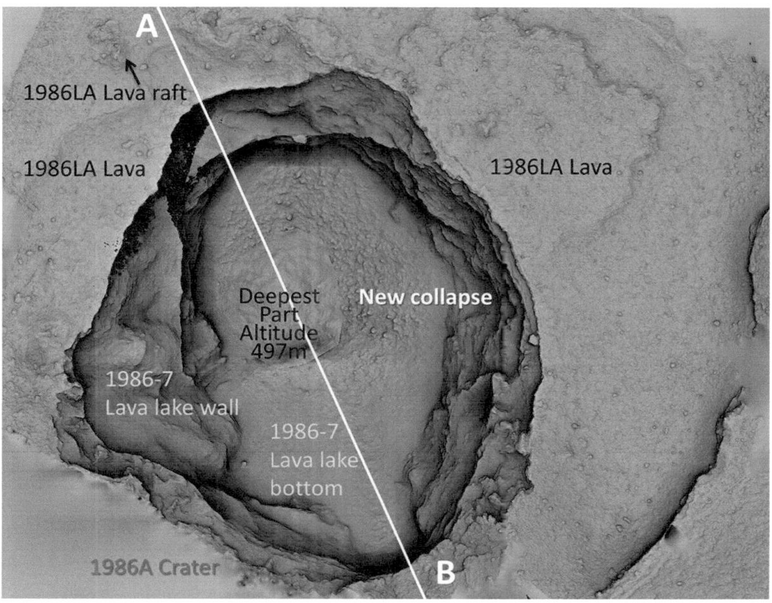

Fig. 9.11 The latest 10 cm size mesh DSM and RRIM of Mt Miharayama central crater

Figure 9.11 shows that Mt. Miharayama central crater bottom was distinguished between the northern part and the southern part. The southern part had traces of the lava lake from the 1986–1987 eruption. In the northern part has no traces of the lava lake. And this part included two parts as the west side and the east side. The west side was getting deeper, and the east side was landfilled with the collapse of the present crater wall. In addition, the overhangs were found on the south-southeast side of Crater A and northwestern side lies behind the lava lake wall of the 1986–1987 eruption.

9.7 Prediction of the Lava Lake Overflows by the 3D Model

Figure 9.12 shows the central crater wall projected on the A-B cross section in Fig. 9.11. It shows the image of the eastern part of the crater wall seen from infinity in the west. The deepest part of the crater bottom was 497 m, and it was largely a sliver type with the crater size increases as it became shallow. We could see further details in the central crater, e.g. the drain back 1987, the bottom sediments, the crater expanded part since 1987, and so on (Fig. 9.12).

We calculated a height and volume curve (H-V curve) by integrating the altitude and the area of the 3D model (Fig. 9.12). Figure 9.12 shows that the lava will overflow at time when the lava lake surface reaches to 685 m of the altitude. The volume of the lava lake at that time would be 10.5 million cubic meters.

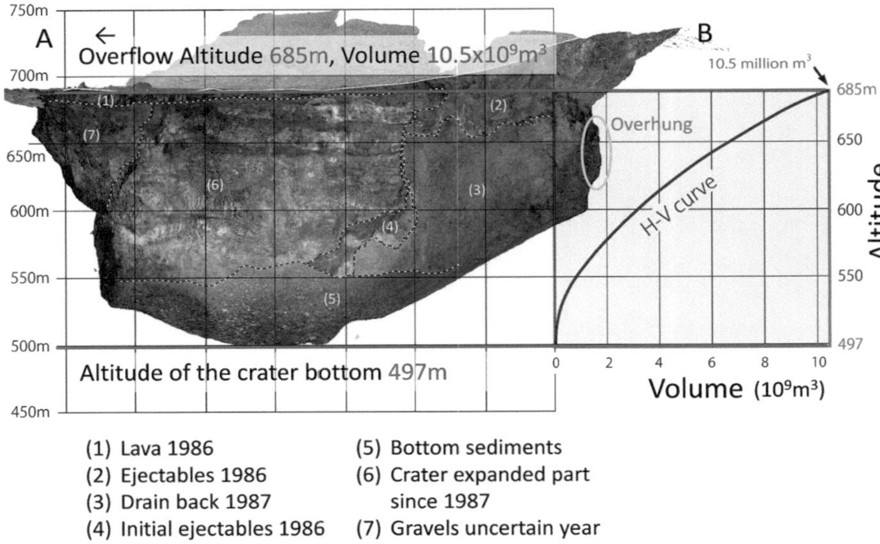

Fig. 9.12 H-V curve of the central crater for prediction of lava lake overflow

9.8 Conclusion

The results show that we can quickly obtain the approximate volume by photographing the lava lake from an easy direction and comparing it with the pattern of crater wall at the eruption. Also, the time of the lava overflow can be predicted by monitoring the increasing height of the lava lake surface. It is an attempt to make a huge measuring cup with the crater. In this study, we confirmed that UAV are effective for active volcano observation and proposed it to be useful for the estimation of lava overflow time.

Acknowledgments We thank the Ministry of Education, Culture, Sports, Science and Technology (MEXT) Japan that provided the opportunity for this study. And we also thank Hokkaido University and Dr. Ram Avtar for inviting us to write this chapter.

References

Chiba T, Urayama T, Mochizuki T, Miura S, Naruke S (2019) Making a 3D model of a crater using UAV. For the future 2019 Asia Air Survey, pp 92–93

Japan Meteorological Agency (JMA). https://www.data.jma.go.jp/svd/vois/data/tokyo/rovdm/Izu-Oshima_rovdm/kazansetumei.html

Oda, K., Hattori, S., Saeki, H., Takayama, T., & Honma, R. (2015) Qualification of point clouds measured by SfM software. The International Archives of Photogrammetry, Remote Sensing and Spatial Information Sciences, 40(4), 125.

Fuse T (2016) SfM and multi view stereo. Photogramm Remote Sens 55(4) 259–262. https://doi.org/10.4287/jsprs.55.259 (in Japanese)

Chapter 10
Applications of UAV Remote Sensing to Topographic and Vegetation Surveys

Hiroyuki Obanawa and Hideaki Shibata

Abstract Three-dimensional models of forestry from Structure from Motion and Multi-View Stereo (SfM-MVS) system have been developed using 2D imagery data captured by UAVs. Such models are useful in estimation of tree height in forestry which is required for several applications of remote sensing. UAV-based SfM measurements allow easy creation of high-resolution ortho-mosaic images and topographic maps. A deep insight into temporal changes in topography and vegetation information is presented. Reconstruction of the volcanic landscapes for the last several years has been carried out using SfM-MVS from ground-controlled aerial photographs. A brief history, development, and prospects of UAV research are also accompanied with research on topography and vegetation using UAV remote sensing.

Keywords UAV · SfM · Remote sensing · Topography · Vegetation

10.1 Introduction

In fields such as environmental science, it is important to accurately grasp the spatial structure and functions of the natural environment and ecosystems, and it is necessary to build and develop observation and analysis technologies to realize them. From around 2013, creating three-dimensional models of objects by Structure from Motion and Multi-View Stereo (SfM-MVS) processing of stereo aerial images taken with a consumer-grade, inexpensive, small-unmanned aerial vehicle (UAV) emerged as a general survey method in numerous research fields (e.g. Obanawa

H. Obanawa (✉)
Hokkaido Agricultural Research Center, National Agriculture and Food Research Organization, Sapporo, Hokkaido, Japan
e-mail: obanawah924@affrc.go.jp

H. Shibata
Field Science Center for Northern Biosphere, Hokkaido University, Sapporo, Hokkaido, Japan
e-mail: shiba@fsc.hokudai.ac.jp

R. Avtar, T. Watanabe (eds.), *Unmanned Aerial Vehicle: Applications in Agriculture and Environment*, https://doi.org/10.1007/978-3-030-27157-2_10

et al. 2014a, b; Secretariat of i-Construction promotion headquarters in Geospatial Information Authority of Japan 2016). Here, we introduce several examples of the use of UAV-measurements to conduct surveys of topography and vegetation which we have contributed to the research literature in recent years.

10.2 Application Examples in Topographic Measurement

Obanawa et al. (2015) conducted UAV-SfM measurements of various environments using only positional information, measured by a single-point positioning Global Navigation Satellite System (GNSS) mounted on a small UAV and verified its usefulness for a topographical survey. As a result, it became clear that this method was not suitable where precise measurements were required, such as accurate measurement of slope inclination and coordinate acquisition of an object, because it included overall vertical displacement and inclination. However, compared with conventional survey methods, there were many advantages such as its application in hard-to-reach areas such as steep slopes, mobility in field surveys, and lower initial and operating costs. UAV-SfM measurements do allow easy creation of high-resolution ortho-mosaic images and topographic maps (Fig. 10.1).

Gomez et al. (2015) applied SfM-MVS to past (1966–2013) aerial photographs to verify the ability of SfM-MVS to gain deeper insight from past topography and vegetation information and examine temporal changes in topography. The survey targets were riverine, coastal, and volcanic topography. As a result, the authors succeeded in recreating three-dimensional geomorphological landscapes of the past and acquiring vegetation information such as tree height. However, due to the low resolution of the early photographs, it was difficult to investigate topographic changes. There was also the difficulty of handling large data sets. In conclusion, SfM-MVS was found to be effective in investigating topographic changes from the middle of the twentieth century to the present, especially in very active geomorphic areas (Fig. 10.2).

Hayakawa et al. (2016) described the basics of SfM-MVS photogrammetry and reviewed its applications in the field of geomorphology to survey slope (mass movement), fluvial, coastal, volcanic, glacial and periglacial, and tectonic topography. The authors summarized a range of studies giving consideration to the coverage area, data resolution, measurement frequency, data accuracy/quality, and the potential for dissemination in the future (Fig. 10.3).

Suganuma et al. (2017) conducted a UAV flight survey in central Dronning Maud Land and the Soya Coast in East Antarctica as part of the survey of the 57th Japan's Antarctic Research Expedition. The authors succeeded in acquiring high-resolution terrain information using UAVs in Antarctica; this required prior training in the use of UAVs in a cold region, taking countermeasure against low temperatures and high winds, and overcoming the UAV position control problem using analysis of GNSS satellite characteristics and making improvements to a GNSS module. As a result, in central Dronning Maud Land a 3D terrain model, allowing detailed polygon topography analysis, could be obtained. In the future, by clarifying the three-dimensional

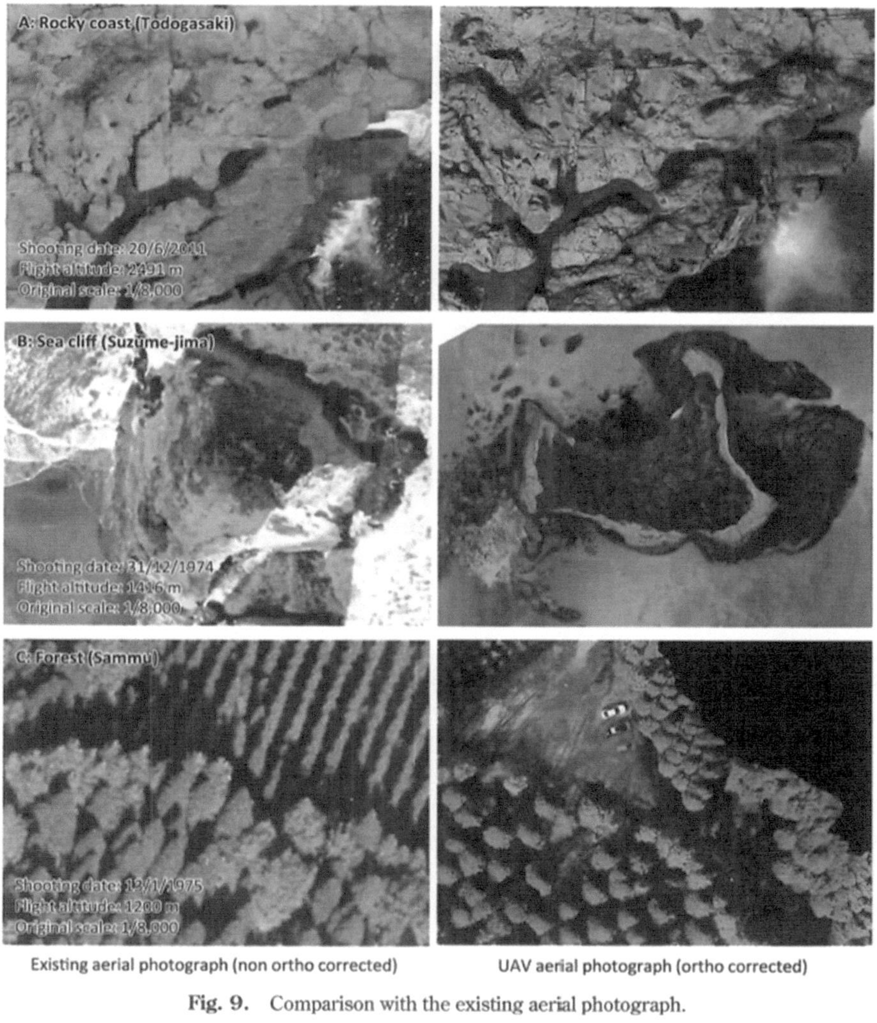

Fig. 10.1 Comparison of existing aerial photograph and UAV aerial photograph. (Adapted from Obanawa et al. 2015)

form, size and distribution characteristics of polygons, there is a good chance of using this data to gain knowledge about formation processes and the relationship of topography with underground structures. On the Soya coast, three-dimensional terrain models of glacial topography and basement rocks, showing characteristic erosion due to long-term weathering, were obtained. In the future, by creating a higher-resolution DEMs, it may become possible to develop this into a quantitative analysis of the topography, for example, using UAV-SfM to verify the weathering

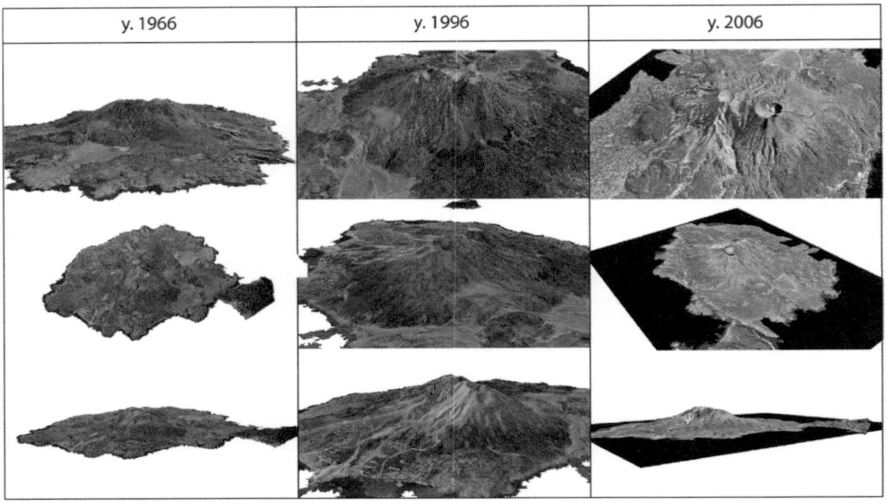

Fig. 10.2 Using ground-controlled aerial photographs from 1966, 1996, and 2006, the SfM-MVS method has allowed the reconstruction of the volcanic landscapes of the last 40 years. The diameter of the volcano is about 10 km. (Adapted from Gomez et al. 2015)

characteristics of basement rock. From the above, the UAV flight survey demonstrated that it is possible to efficiently and extensively conduct an Antarctic survey where the investigation period is limited and that these methods can overcome many limitations of conventional field surveys (Fig. 10.4).

Hayakawa et al. (2018) investigated hummocks on the northeastern flank of Mt. Erciyes in Kayseri, central Turkey using UAV and SfM-MVS. Although high-resolution aerial images or satellite images were limited, new knowledge on the formation of hummocks and debris avalanche deposits (DADs) was obtained by rapid on-site observation using the UAV-SfM method. According to detailed topographical data, hummocks were arranged along the flow direction of a debris avalanche and showed that an extensional regime was dominant; some displaced hummocks were also distributed, however, indicating a compressional regime also contributed. These topographical analyzes suggested that the behavior of a debris avalanche and the formation of hummocks are subject to existing terrain, especially the former caldera wall. These findings should be useful for future disaster prevention plans (Fig. 10.5).

Saito et al. (2018) used UAV and SfM-MVS to examine the topographical features of the landslide caused by the 2016 Kumamoto earthquake (Mw 7.1) in the Sensuikyo area (1.0 km^2) of the Aso volcano, Japan. Landslides due to heavy rainfall (e.g., in 1990, 2001, and 2012) frequently occurred in the investigation area. The authors created ortho-mosaic images and DSMs with a resolution of 0.06 m before and after the 2016 Kumamoto earthquake. Using these high-resolution data, a total of 54 seismic landslides with a volume of 9.1–3994.6 m^3 were extracted. Many of these landslides were located along the ridge and occurred in the upper part of the

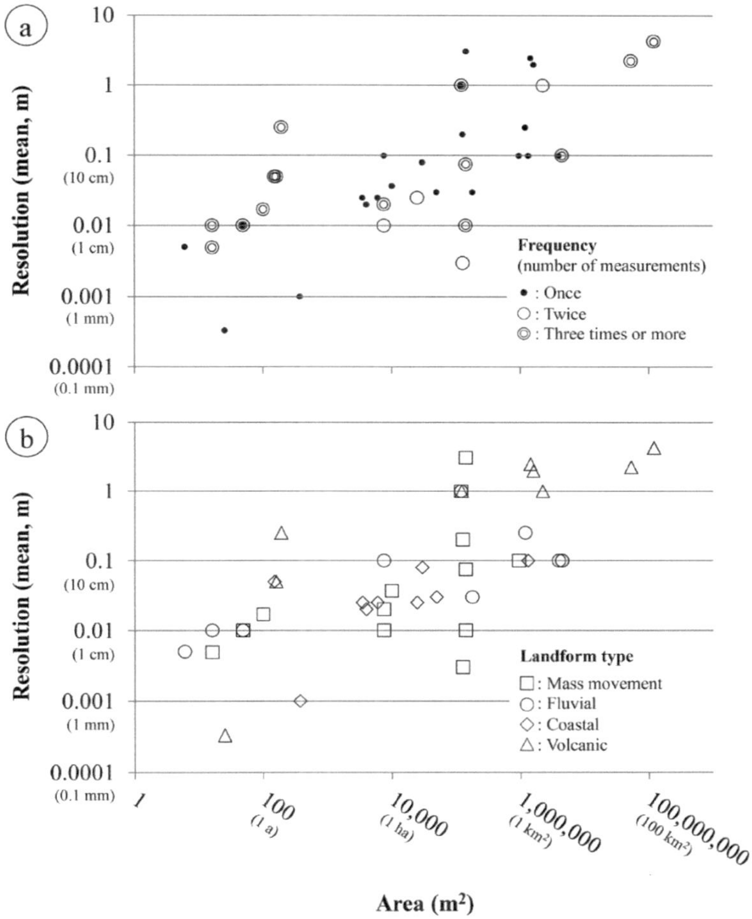

Fig. 10.3 Relationships between measurement area and spatial resolution of data produced by SfM-MVS in papers on geomorphological studies. The plots are classified by (**a**) temporal frequency or number of measurements and (**b**) type of landforms targeted. (Adapted from Hayakawa et al. 2016)

landslide scars formed during heavy rainfall in 2012; this suggests that the topographic effect on earthquake shaking, i.e., the amplification of ground surface acceleration, greatly affected the occurrence of landslides. The average depth of the earthquake-induced landslide was 1.5 m, which was deeper than the rainfall-induced landslide. The total sediment production of the earthquake-induced landslides reached 2.5×10^4 m^3/km^2, which was the same order of magnitude as previous rainfall-induced landslides. Therefore, the effect on sediment production and topography change of 2016 Kumamoto earthquake was equivalent to the previous heavy rain-induced landslide (Fig. 10.6).

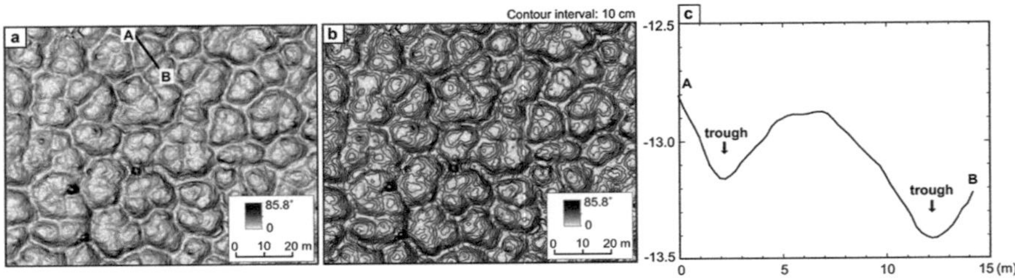

Fig. 10.4 Polygon structure obtained with the UAV-SfM analysis as a (**a**) slope gradation map, (**b**) contour map with 10-cm intervals, and (**c**) cross-section of the polygon. (Adapted from Suganuma et al. 2017)

Obanawa and Hayakawa (2018) surveyed a sea cliff of a Peninsular rock, named Suzumejima Island, located in eastern Chiba prefecture, Japan, using UAV and terrestrial laser scanning (TLS), and evaluated the volumetric changes in the cliffs with a high spatial resolution (centimeters). Seven measurements were carried out over 2 years, approximately every 4 months. It was impossible to survey the whole area from the ground because the sea surrounded the island on three sides and the surrounding slopes of the island were very steep. Surveying was also made difficult by the complicated rock form, including reverse gradient (overhang), which made measurements from high altitude satellites or Cessna aircraft impossible. The authors succeeded in acquiring three-dimensional data with high accuracy and high resolution, from a target that cannot be measured using existing methods, by merging measurement data from UAV and TLS. It was possible to photograph and determine the shape of the vertical and overhanging cliff faces by changing the shooting direction of the camera mounted on the UAV. In addition, since the running cost and the labor involved in taking the measurements were small, it was easy to collect multi-temporal data at a high frequency which could be used to elucidate details of erosion mechanisms. The authors quantified the amount of geomorphological change (erosion amount) in each measurement period through analysis of three-dimensional point cloud data to calculate the changes in the rate of erosion over time. Furthermore, the authors examined strong earthquakes and high sea waves as potential causes of erosion by considering the correspondence between the occurrence of each event and the rate of erosion during each measurement period. The results predict that the sea waves were more influential to the erosion of the cliff than earthquakes (Fig. 10.7).

As other special cases of topographic measurement, Obanawa et al. (2016) used the UAV-SfM method to estimate the snow depth distribution in a mountainous area and demonstrated the effectiveness of the method by comparing it with the actual measurement value. Moreover, Murakami et al. (2016) and Obanawa et al. (in print) tried three-dimensional measurement of coral reef in the sea and showed its effectiveness by comparing these measurements with the observed values from traditional survey techniques.

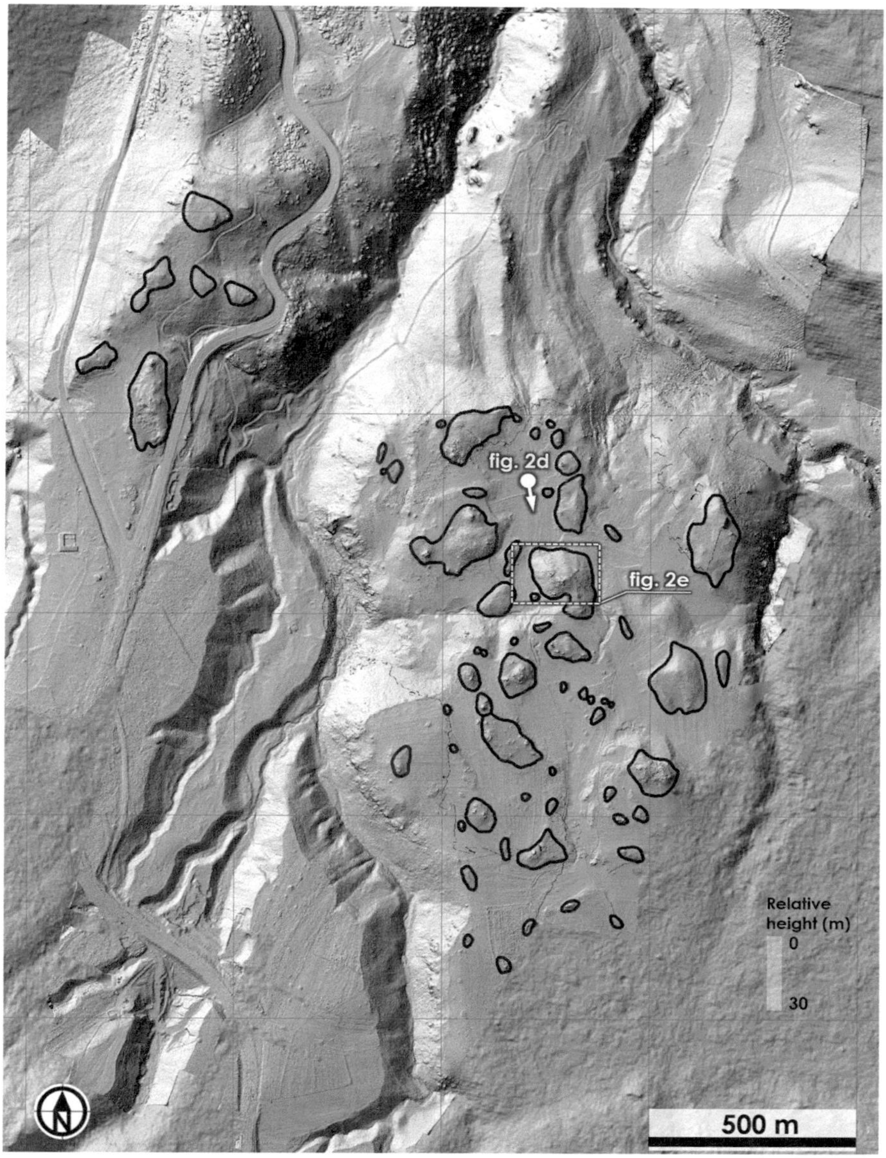

Fig. 10.5 Map of the hummocks identified in the study area ($n = 65$). Blue to purple colors show the sink-filled areas on the inverse DEM, indicating the locations of local mounds and ridges. Black solid lines represent the mound boundaries of hummocks identified. Note that not all of the sink-filled areas are identified as hummocks; this is based on the interpretation of orthorectified images and field observations. Hillshade images are derived from the RPAS-derived 36 cm DEM (foreground) and 10 m PRISM DEM (background). (Adapted from Hayakawa et al. 2018)

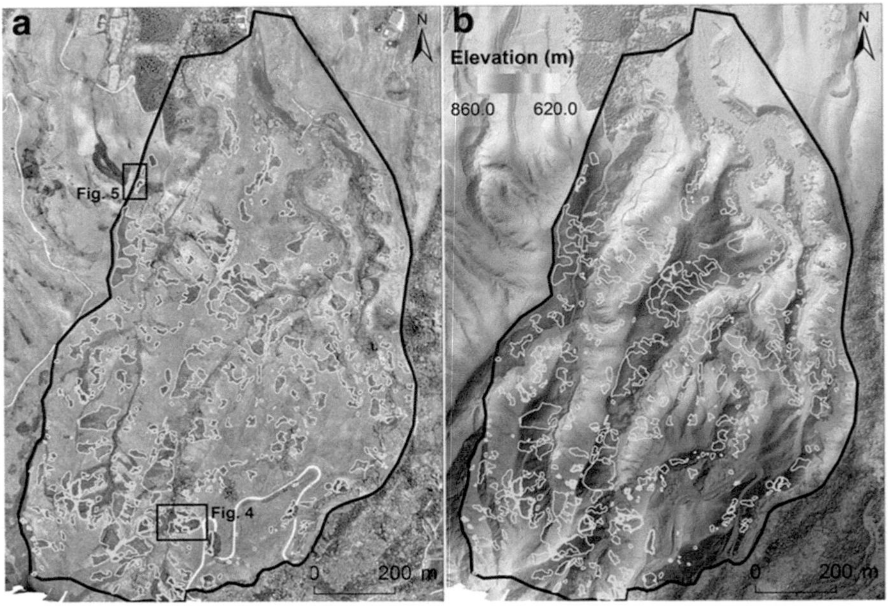

Fig. 10.6 Distributions of landslide initiation areas triggered by the 2016 Kumamoto earthquake (yellow polygons) and the heavy rainfalls of 2012 (white polygons) with the (**a**) orthorectified images and (**b**) digital surface models (DSMs), with a spatial resolution of 0.06 m, acquired in May 2016. The black polygon shows the study area. (Adapted from Saito et al. 2018)

10.3 Application Examples in Vegetation Measurement

Tamura et al. (2015) carried out aerial photography with a UAV a total of three times on a relatively flat area and a slope, in a deciduous broad-leaved forest and a coniferous mixed forest, respectively, in Uryu experimental forest of Hokkaido University, Japan. A three-dimensional model, Digital Surface Model (DSM), Digital Terrain Model (DTM), and Digital Canopy Model (DCM) were created from the aerial images. Treetops were extracted from the DCM and compared with the actual values obtained in a field survey. As a result, it was found that seasonal change affects tree height measurement and the accuracy of tree height measurement by SfM was 1.40 m (RMSE) for deciduous trees and 1.48 m for conifers before defoliation (Fig. 10.8).

Iwata et al. (2018) developed a precision forestry technology to increase the amount of cellulose per unit area of trees in a Eucalyptus plantation in northern Brazil. As part of that, the authors developed a technology to monitor the amount of large-area biomass more efficiently and with high accuracy by acquiring and analyzing three-dimensional data of the forest using a ground 3D laser scanner and a UAV (Fig. 10.9).

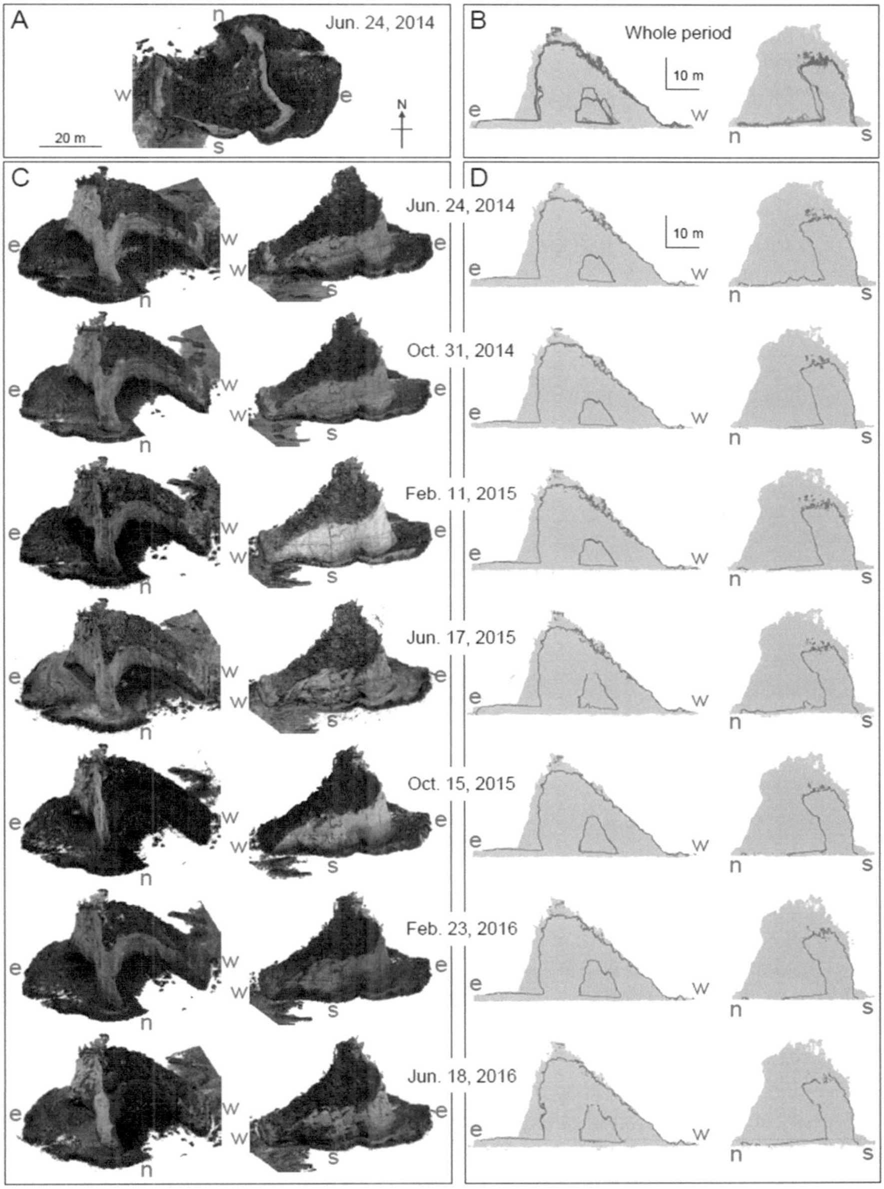

Fig. 10.7 3D models (**a**, **c**) and cross-sections (**b**, **d**) of Suzumejima from June 24, 2014, to June 18, 2016. **a** is viewed from the top and (**c**) is viewed obliquely from above with traverse lines (red lines) of cross-sections. Lowercase letters (n, s, e, and w) indicate the directions of traverse lines. The left figures of (**c**) are viewed from the north, and the right figures are viewed from the south. **b** shows superimposed cross-sections of all periods (**d**). The left figures of **d** show cross-sections of an east-west orientation, and the right figures show cross-sections of a north-south orientation. (Adapted from Obanawa and Hayakawa 2018)

Fig. 10.8 Treetops extracted from Digital Canopy Model. (Adapted from Tamura et al. 2015)

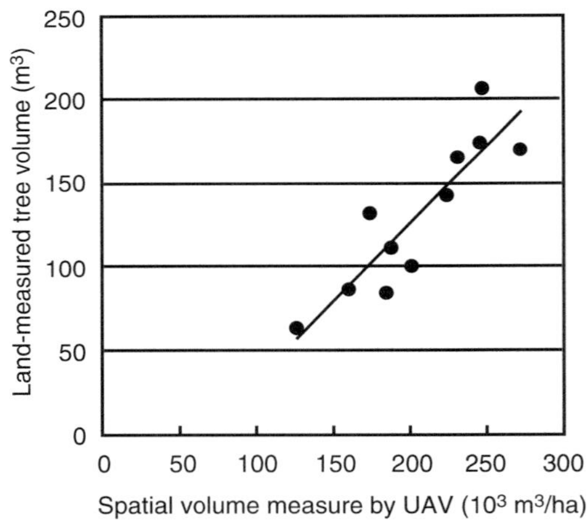

Fig. 10.9 Correlation of UAV data and land-measured values. (Adapted from Iwata et al. 2018)

Inoue et al. (2019) used the UAV-SfM method to estimate the spatial distribution of leaf nitrogen content on the surface of upper trees and understory vegetation (*Sasa*) in forests with heterogeneous tree species composition and canopy structure. The survey site was a natural cool-temperate mixed forest in Uryu experimental forest of Hokkaido University located in northern Hokkaido, Japan. First, the relationship between the nitrogen concentration in leaves of tree and *Sasa* and the RGB index value was determined. Next, the canopy height was estimated from the difference between the surface height of the canopy and the ground height. The authors also developed a method to create a spatial distribution map of evergreen conifers, deciduous broad-leaved trees, and *Sasa* using the difference between the surface height of the canopy and the defoliation and evergreen indices. Finally, the spatial distribution of nitrogen concentration in canopy leaf at the basin scale was estimated using the, newly created, vegetation distribution map and the relationship between the RGB index and nitrogen (Fig. 10.10).

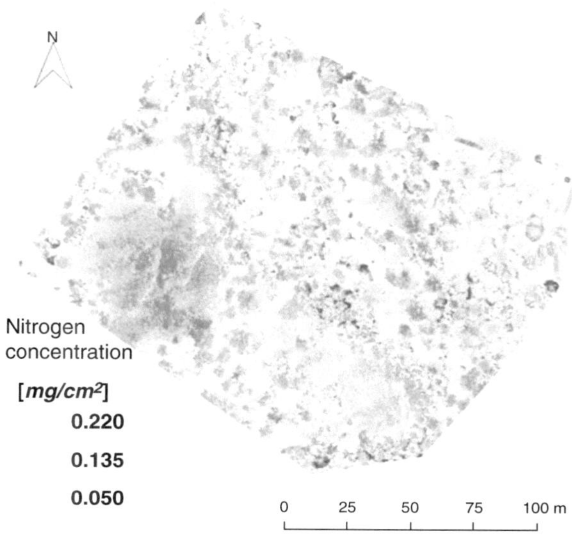

Fig. 10.10 Estimated spatial distribution map of nitrogen concentration per leaf area included in the canopy. (Adapted from Inoue et al. 2019)

10.4 Conclusions

A surveying method using consumer-grade, low-cost UAV and SfM-MVS technologies, which emerged in 2013, initially led to many research reports assessing how best to use the equipment and the accuracy of the survey results. Following initial testing, higher-resolution measurements of various landforms that had been previously difficult to measure were performed, and quantification of topographical changes with higher resolution in space and time was achieved by comparing data over multiple periods. In forestry, UAV and SfM-MVS technologies have enabled tree height measurements, with higher resolution than previously possible, to be performed and have been used to estimate the biomass and nitrogen content of leaves. Furthermore, as a derivative usage, SfM-MVS technology has been used for surveying coral reefs in marine environments and measuring snow depth distribution in mountainous areas.

Development in UAV-SfM technology will be achieved through the improvement of hardware that will improve the UAV function (for example, strengthening the waterproof/windproof function or extending the flight time by using a next-generation battery), image quality of cameras, and calculation speed of the computer performing SfM-MVS processing. These developments will likely contribute to the continued expansion of the application range of UAV-SfM technology.

References

Gomez C, Hayakawa YS, Obanawa H (2015) A study of Japanese landscapes using structure from motion derived DSMs and DEMs based on historical aerial photographs: new opportunities for vegetation monitoring and diachronic geomorphology. Geomorphology 242:11–20. https://doi.org/10.1016/j.geomorph.2015.02.021

Hayakawa YS, Obanawa H, Saito H, Uchiyama S (2016) Geomorphological applications of structure-from-motion multi-view stereo photogrammetry: a review. Trans Jpn Geomorphol Union 37(3):321–343

Hayakawa YS, Yoshida H, Obanawa H, Naruhashi R, Okumura K, Zaiki M, Kontani R (2018) Characteristics of debris avalanche deposits inferred from source volume estimate and hummock morphology around Mt. Erciyes, Central Turkey. Nat Hazards Earth Syst Sci 18:429–444. https://doi.org/10.5194/nhess-18-429-2018

Inoue K, Shibata H, Yoshida T, Nakaji T, Obanawa H, Kato A (2019) Spatial assessment of leaf nitrogen content in a natural cool-temperate mixed forest using unmanned aerial vehicle derived 3D data. Jpn J For Environ 61(1):1–13. https://doi.org/10.18922/jjfe.61.1_1

Iwata E, Suyama K, Urata N, Nakahama K, Nanto K, Shinya T, Kawaoka A, Shibusawa S, Kodaira M, Kato A, Obanawa H (2018) Precision forestry technology in industrial plantation. Jpn TAPPI J 72(7):33–38

Murakami T, Obanawa H, Kohno H, Shimokawa S, Tabayashi Y, Mizutani A (2016) Applicability of underwater 3D measurements by SfM for coral reef waters. Annu J Civil Eng Ocean JSCE 72 (2):766–771. https://doi.org/10.2208/jscejoe.72.I_766

Obanawa H, Hayakawa YS (2018) Variations in volumetric erosion rates of bedrock cliffs on a small inaccessible coastal island determined using measurements by an unmanned aerial vehicle with structure-from-motion and terrestrial laser scanning. Prog Earth Planet Sci 5:1–10. https://doi.org/10.1186/s40645-018-0191-8

Obanawa H, Hayakawa YS, Gomez C (2014a) 3D modelling of inaccessible areas using UAV-based aerial photography and structure from motion. Trans Jpn Geomorphol Union 35 (3):283–294

Obanawa H, Hayakawa YS, Saito H, Gomez C (2014b) Comparison of DSMs derived from UAV-SfM method and terrestrial laser scanning. J Japan Soc Photogram 53(2):67–74. https://doi.org/10.4287/jsprs.53.67

Obanawa H, Hayakawa YS, Kato A, Gomez C (2015) Simplified survey method using small UAV and single-point positioning GNSS equipped digital camera. Trans Jpn Geomorphol Union 36 (2):87–106

Obanawa H, Kawashima K, Matsumoto T, Iyobe T, Ohmae H (2016) Measurement of snow distribution using small UAV. Seppyo 78(5):317–328

Obanawa H, Tabayashi Y, Murakami T, Kohno H, Shimokawa S, Mizutani A (in print) Underwater three-dimensional measurements. In: Shimokawa S, Murakami T, Kohno H (eds) Geophysical approach to marine coastal ecology – the case of Iriomote Island, Japan. Springer, Tokyo

Saito H, Uchiyama S, Hayakawa YS, Obanawa H (2018) Landslides triggered by an earthquake and heavy rainfalls at Aso volcano, Japan, detected by UAS and SfM-MVS photogrammetry. Prog Earth Planet Sci 5:1–10. https://doi.org/10.1186/s40645-018-0169-6

Secretariat of i-Construction promotion headquarters in Geospatial Information Authority of Japan (2016) Outline of operating manual and safety guidance for UAV survey on the public survey. J Japan Soc Photogram 55(3):210–216. https://doi.org/10.4287/jsprs.55.210

Suganuma Y, Kawamata M, Shiramizu K, Koyama T, Doi K, Kaneda H, Aoyama Y, Hayakawa H, Obanawa H (2017) Unmanned aerial vehicle (UAV)-based survey in Antarctica for high-definition topographic measurements. J Geogr 126(1):1–24. https://doi.org/10.5026/jgeography.126.1

Tamura T, Kato A, Obanawa H, Yoshida T (2015) Tree height measurement from aerial images taken by a small unmanned aerial vehicle using structure from motion. J Jpn Soc Reveget Tech 41(1):163–168. https://doi.org/10.7211/jjsrt.41.163

Chapter 11
The Role of Infrared Thermal Imaging in Road Patrolling Using Unmanned Aerial Vehicles

Neha Sharma, A. S. Arora, Ajay Pal Singh, and Jaspreet Singh

Abstract In the past few years, the tremendous growth in road network and vehicles has increased the road fatalities at a very alarming rate. Road patrolling is one of the prominent measures to reduce road fatalities. Generally, road patrolling has been done using manned ground vehicles whose performance is highly dependent on environmental conditions. With this in mind, an infrared (IR) thermal imaging-based technique to enhance the object's detection in poor weather conditions is presented in this study. Moreover, it can be employed in unmanned aerial vehicles (UAVs) for road patrolling in unfavorable weather conditions including total darkness, fog, and heavy rain. The aim of this study is to automate the process of object detection which enhances road patrolling, where it can enforce the traffic safety compliances and provide automatic rescue call facilities in case of remote area fatalities. The proposed approach is comprised of three steps: (a) data acquisition, a dataset of 53 thermograms at various weather conditions has been created; (b) data processing, a thresholding method, morphological operations, and pseudo-coloring have been performed; and (c) results validation, compare the outcomes of proposed methodology with standard approaches. More specifically, the optimal temperature thresholding in conjunction with morphological operations automates the process of object detection, where the pseudo-coloring algorithm is introduced to convert the thermograms into RGB space which enhances the images for better visualization. Consequently, the proposed methodology shows a good accuracy of 83% for object detection in different weather conditions. The methodology can be used with UAVs which enables fast monitoring of recent accidents on remote locations as the clashing of vehicles raises the temperature. Besides, the issues and challenges faced in the thermal-based UAVs are also discussed.

N. Sharma · A. P. Singh
Department of Electronics and Communication Engineering, Sant Longowal Institute of Engineering & Technology, Longowal, Sangrur, Punjab, India

A. S. Arora (✉) · J. Singh
Department of Electrical & Instrumentation Engineering, Sant Longowal Institute of Engineering & Technology, Longowal, Sangrur, Punjab, India

R. Avtar, T. Watanabe (eds.), *Unmanned Aerial Vehicle: Applications in Agriculture and Environment*, https://doi.org/10.1007/978-3-030-27157-2_11

Keywords Thermal imaging · UAVs · Night vision system · Optimal temperature thresholding · Pseudo-coloring

11.1 Introduction

The foremost mode of transport in every nation is the road transport whether it is the transportation of goods or people. To reach the demand for road transport, the length of road network and the number of vehicles are increasing over the last few decades. As a consequence, evolution in road network leads to a negative impact, i.e., with increase in urbanization and motorization in the country, there is a tremendous increase in road accidents and road crash fatalities. Today, injuries due to road traffic are one of the prominent causes of death, hospitalization and disabilities in the country which results in huge socio-economic costs.

Thus, tremendous increase in number of vehicles on roads has caused an increase in accidents at a very alarming rate. In India, it has been analyzed that the fatalities due to road accidents have increased by 5% every year during the last 10 years (Road Crash Statistics 2016). As reported by Ministry of Road Transport and Highways of India, a total of 480,652 road crashes happened in 2016, in which 150,785 people lost their lives and 494,624 were injured. Statistically, 413 lives have been lost per day in 1317 crashes and 17 lives have been lost in 55 crashes every hour. Furthermore, from 2007 to 2017, this figure of road fatalities has increased by 31%, whereas 25.6% increase was observed in fatal road crashes in the same period (Singh 2017). The major and growing public health problem in India is the injuries and fatalities resulting from road traffic accidents.

Road accidents induce a major loss of life, economic growth, and productivity, both at the personal and professional levels. Besides, road accidents increase the energy consumption and travel time which further disturb the routine activities. With the technological advancement in automobiles, the modern vehicles are more fuel efficient, inexpensive, safe, and automated. Thus, people prefer their own vehicle instead of public transport which increases the chances of road accidents due to heavy rush of vehicles on roads. However, accidents prevention systems are available in modern vehicles, but the fatalities due to road accidents are increasing day by day which have become a great challenge to the nation and public health agencies (Singh 2017). There are many reasons for road accidents, such as overloading of vehicles, poor construction, distracted driving, drunken driving, disobeying traffic rules, and ignorance of helmets, seat-belts, and child restraints usage. Mostly, the youngsters are losing their lives because of rash driving, drunken driving, and over speeding which is a great loss to the nation. In many situations, human lives are lost due to poor communication, delayed medical assistance, poor environmental conditions, and heavy traffic on the road.

11.1.1 Road Accidents

As per the records reported by World Bank, it is highlighted that if the death rate caused by road accidents was halved, it could increase the GDP by 7 to 22% per capita over the 24 years in selected countries. For a country like India, where there is no formal social welfare support for a vast majority of the population belonging to lower and middle-class families, a reduction in road fatalities would give rise to a progressive phenomenon (Singh 2017). Among the main factors behind the road accidents, over-speeding is the major factor which caused 70.4% of all accidents, in which 66.7% individuals died and 72.8% individuals were injured.

11.1.1.1 Major Causes

Road accidents have multi-causal factors, considering the errors due to humans, defects in roads, vehicle defects, non-availability of pedestrian facility, factors due to circumstances such as poor weather conditions, and wandering of stray animals. Some major causes of road accidents are:

- ***Over speeding and overtaking***: A vehicle moving with high speed has more chance to cause accident to self and other vehicles.
- ***Drunk driving***: Alcohol and drugs intake by drivers reduces their concentration and cause accident which proves fatal.
- ***Distracted driving***: Distraction of drivers due to mobile phone usage, daydreaming, eating, and talking with occupants while driving may cause road accidents.
- ***Unfavorable driving conditions***: The visibility impairment due to unfavorable driving conditions, such as fog, rain, hailstorm, and complete darkness can also be the factors for road accidents.

Due to better visibility of vehicles and roads in clear weather, it is more likely that the drivers speed up the vehicle as compared to rainy and foggy conditions. Meanwhile, in poor visibility conditions, drivers experience a higher Standard Deviation of Lane Position (SDLP) (i.e., poor lane keeping) due to low visibility of road surface markings (Das et al. 2019). Nowadays, there is progressive increase in the development of driver assistance and accident prevention systems which modify the vehicles for comfortable and safe driving.

11.1.1.2 Precautionary Measures

From various studies, it has been clear that the driving conditions are becoming life threatening day by day. Therefore, some methods should be carried out to have control over such fatalities. Some of the measures to reduce fatalities are enlisted as:

- ***Accident prevention system***: With the technological advancement, the accident prevention systems, such as air bags, anti-lock braking system (ABS), etc. are equipped in modern vehicles.
- ***Night vision system (NVS)***: Poor visibility at night, in fog and smoke conditions put pedestrians at risk, as objects at farther distance are usually not seen by drivers. An NVS uses thermal cameras and can see objects beyond the visible light. Therefore, it becomes possible to detect any obstacle in poor visibility conditions, like darkness, fog, heavy rain, hailstorm, etc. NVS can be referred as the system that enhances the vision of a driver in unfavorable driving conditions.
- ***Road patrolling***: Usually, the manned ground units are employed to provide fast surveillance on highways and to cover remote locations. But, accidents have no pre-defined place which makes road patrolling a tough task with the manned units. Besides these limitations, their performance can also be affected with the external environmental factors like night, fog, smoke, and heavy rain. So, it is noteworthy to say that there is a huge requirement of such unmanned aerial vehicles (UAVs) equipped with such system which is not influenced by above-discussed factors.

11.1.2 Infrared Radiations (IR) and Thermal Imaging

Sir William Herschel, a British astronomer and musician, discovered the IR rays in 1800 (Gade and Moeslund 2014). He found that there is a type of invisible radiation that lies beyond the red band of electromagnetic (EM) spectrum which also gave rise in temperature on thermometer (Gade and Moeslund 2014). IR radiations have wavelength shorter than those of microwave band and longer than those of visible band of EM spectrum and therefore not visible to the naked human eyes.

These radiations cover the wavelength band of 0.75 – 1000 μm. Due to molecular vibration, every object which has temperature above absolute zero (−273.15 °C) radiates IR energy, even cold objects (Singh and Arora 2018). The energy of emanated IR radiations is a function of surface temperature, characteristics of object material, and surface. In brief, IR radiation is a function of the surface temperature of an object, thus the temperature of an object can be measured without having physical contact with the object using IR radiation. There are various devices which can detect IR radiations, like bolometer, thermopile, etc. which are sensitive to heat.

IR thermography is a thermal radiometric approach which is comprised of (a) measurement of IR radiation emanating from an object's surface; (b) converts these radiations into thermal data; (c) represent it in the form of 2D false color images (or thermograms as shown in Fig. 11.1) for visual perception; and (d) image processing and data analysis (Singh and Arora 2018). Thermal imaging technique has become popular due to its non-contact and non-invasive nature. It converts the invisible radiations of any object into visible images for analysis. IR thermal imaging is applicable in all fields where the temperature differences can be used for the analysis of a process. Firstly, it was developed for military purposes but subsequently gained its interest in medical, food, agriculture, civil engineering, aerospace, and many more fields.

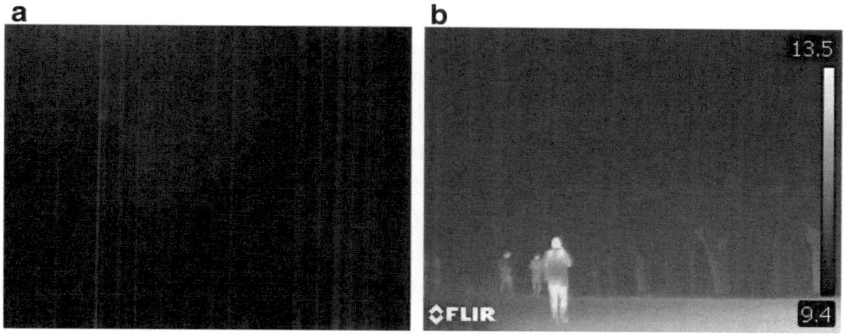

Fig. 11.1 Illustrates the people on road in darkness (**a**) Visual image (**b**) Thermal image

As discussed, an NVS is one of the prominent precautionary measures which assist the driver in poor weather conditions. In NVS, the thermal camera or IR camera has been employed. Usually, an NVS is classified as active and passive, where an object of interest is excited with an IR source in case of active technique (Piniarski et al. 2014). The active NVS has wavelength of near IR light (0.7–1.1 μm) close to the visible range. A powerful IR light source is required which emits radiations on surrounding objects and then these IR radiations are reflected back towards the sensors to create an image. However, it exhibits low cost, small size, and provides high resolution image, but it cannot work well in poor weather conditions. In case of passive system, the heat radiated by an object is captured for imaging without any external excitation source. The detection range is 300 m which is larger than active systems and it works well in poor weather conditions.

11.1.3 Unmanned Aerial Vehicle

A UAV or the unmanned aircraft system can be operated in various modes, such as manually controlled, autonomous, semi-autonomous, or can be used in combination (Austin 2011). These are capable of performing the tasks efficiently and as many as humans can think off. In spite of the fact that UAVs were mainly designed for military applications, the UAVs have successfully been used on scientific, general public safety, and certain commercial tasks such as image acquisition of catastrophic places, constructive buildings, communication relays, rescue from danger, traffic surveillance, border surveillance, industrial inspection, pesticides spraying, and so on.

Nowadays, in busy and complex environment, there is need of UAVs for various applications like fast surveillance, accessibility to remote locations, aerial monitoring, and many more, which cannot be performed by manned and ground vehicles. The different types of imaging systems are employed in UAVs, where the type of imaging is highly reliant on applications. But, IR thermal imaging exhibits different features which cannot be obtained with other imaging modalities. Also, the technological advancements in various fields enforce its use almost everywhere, where

UAVs are not exceptional. So far, the role of thermal camera-based UAVs has been used in various studies such as inspection of photovoltaic plants (Quater et al. 2014), wildfire detection (Riggan and Hoffman 2003; Gonzalez et al. 2016), roof insulation (Zhang et al. 2015), geothermal imaging (Nishar et al. 2016), vegetation monitoring (Berni et al. 2009), remote sensing (Everaerts 2008), and many more.

Besides, the UAVs equipped with thermal imaging cameras can be used for road patrolling which facilitates many tasks such as frequent accident monitoring in remote locations, night vision of road traffic, and visual enhancement in poor weather conditions. With this consideration, the automated approach to detect objects with thermal imaging in poor weather conditions is proposed in this study.

11.2 Radiometric Thermal UAVs

A radiometric thermal drone or UAV is a device which makes use of radiometric functionality of the thermal camera to capture the aerial view by interpreting the intensities of IR signals coming from an object's surface. Radiometric thermal imaging exhibits unique characteristics which cannot be exhibited by visual imaging systems, such as provides heat signatures, enhances visibility in poor weather, and functioning of object based on thermal variations. Due to these properties, radiometric thermal UAVs are highly preferable for security and protection tasks, such as capturing the catastrophic events, border surveillance, industrial inspection, rescuing, and so on. The performance of UAVs can be characterized by (a) imaging quality, which is mainly based on IR detector and optic lens; and (b) flight efficiency, which depends upon the various factors. The UAVs design has three major limiting factors which must be minimized for better flight performance. These factors are size, weight, and power consumption (acronym SWaP).

11.2.1 Characteristics of Radiometric Thermal UAVs

As mentioned, the imaging performance is highly dependent on IR detectors and optic lens. Thus, the IR detectors of higher resolution with higher quality of lens are essential for accurate imaging of aerial view. In addition, to capture the view at high altitude, the optic lens must have long focal length. Meanwhile, the camera outer body and lens must be firm to withstand at unfavorable environmental conditions.

Besides, for efficient flight, the system must meet the criteria of SWaP. UAVs payload is one of the vital characteristics which defines the weight carrying capacity during the flight. For heavy payload UAVs, a strong propulsion system is required which in return makes the system less fuel efficient. Thus, to minimize the power consumption (maximize the flight time), the size and weight of a UAV must be reduced.

11.2.2 Challenges and Issues

Technical As the imaging performance cannot be compromised in UAVs (especially in security arena), the IR detectors and lenses of large sizes are employed which increase the system weight and hence SWaP goal becomes hard to achieve.

Environmental Remote measurement of temperature relies on the ability of the camera to compensate the interferences due to atmosphere, surface characteristics (emissivity and reflectivity) and IR detector itself. The atmosphere is constituted by various gases which act as a filter (atmospheric absorption) to the IR radiations and hence affect the imaging performance.

11.3 Related Work

There are numerous methods which have already been proposed for detection of objects using IR thermal imaging. Some methods used Otsu and local thresholding techniques (Saito et al. 2008) to get better segmentation in night time. In some studies, the efficient use and calibration of night vision cameras (both active and passive) have been discussed (Tsuji et al. 2002; Omar and Zhou 2007; Luo et al. 2010; Piniarski et al. 2014). Other related works used filtering and background subtraction techniques (Rajkumar and Mouli 2015). Also, the experiments have been done by fusing results of visible imaging with thermal imaging which showed better outcomes with respect to the other standard techniques (Goubet et al. 2006). Another approach based on morphological characteristics of human shape which is used to identify the pedestrians is presented in literature (Bertozzi et al. 2004). A review on morphology has been discussed to get wide spectrum about these operations (Arora and Pandey 2016; Pandey and Mathurkar 2017). Some methods which restore the image colors have also been taken into account (Singh and Arora 2017; Jayadevan and Navas 2014). However, most of the methods depend on the assumption that the objects are hotter than the background which is not always applicable. Some of the methods did not give any idea about the detection in odd environmental conditions.

So far, the role of thermal imaging-based UAVs in road patrolling has not been discussed in any study. In this study, the automatic approach to detect the objects in poor weather conditions has been proposed. Also, the methodology can be used with UAVs which enables the fast monitoring of recently happened accidents on remote locations as the clashing of vehicles raises the temperature.

11.4 Proposed Methodology

In this paper, our aim was to automate the process of object detection in poor weather conditions, such as total darkness, rain, and fog using IR thermal camera. The proposed methodology is comprised of three steps: (a) data acquisition, a dataset of 53 thermograms at various weather conditions has been created; (b) data processing, a thresholding method, morphological operations and pseudo-coloring have been performed; and (c) results validation, compare the outcomes of proposed methodology with standard approaches. More specifically, the approach is based on optimal temperature thresholding followed by morphological operations which enhances the automatic detection of objects based on thermal patterns. Then, pseudo-coloring algorithm is introduced to convert the thermograms into RGB space which further enhances the region of interest for better visualization. An optimal thresholding technique is employed to efficiently remove the background interferences which lead to clear segmentation outcomes. Furthermore, the morphological operations are applied which provide clear outlines of the detected target.

11.4.1 Dataset

The FLIR E60 thermal imaging camera has been used to obtain the thermograms at various weather conditions, such as fog, rain, and total darkness. It produces the radiometric thermal data of 320×240 pixels at $\pm 2^{\circ}$C accuracy with $25^{\circ} \times 19^{\circ}$ field of view and has thermal sensitivity less than $0.05^{\circ}C$. In this study, the thermograms were taken at different places with different angles which give high variability in dataset. Also, the camera settings were kept constant during the data acquisition. The dataset of total 53 thermograms has been created for this study. The distance of objects taken was about 10–50 m. The FLIR® tools and MATLAB software were used to interface the computer with thermal camera while data acquisition.

11.4.2 Optimal Temperature Thresholding

To remove the background interferences, the image pre-processing has been performed which characterizes the thermogram with black and white, where the white regions indicate the region of interest and black region indicates the background. In brief, the optimal thresholding converts the complex pattern color thermograms into binary image which provides ease in the automatic detection of objects by removing the background, as shown in Fig. 11.2 (b). However, the Otsu's thresholding is a standard approach for image segmentation, but it didn't provide satisfactory results for this dataset. An iterative method of optimal threshold has been applied which converts the gray level image into binary, as proposed by Singh and Arora (2018). To get the optimal threshold value, the procedure is as follows:

Algorithm: Optimal temperature thresholding

Require: Thermogram I

Require: Set probability of background and foreground as $P_\beta = 0.4$ and $P_\gamma = 0.6$ respectively

Require: Set maximum number of iterations $N = 100$ and tolerance $T_{ol} = 0.001$

Step 1: Calculate the maximum temperature T_{max}, minimum temperature T_{min} and variance σ^2 of thermogram I

Step 2: Initial temperature threshold, $T_\cap^1 = \frac{(T_{max} + T_{min})}{2}$

Step 3: **for** i = 1 **to** N **do**

Step 4: Background image, $I_\beta^i = \begin{cases} 1, & I < T_\cap^i \\ 0, & else \end{cases}$

Foreground image, $I_\gamma^i = \begin{cases} 1, & I \geq T_\cap^i \\ 0, & else \end{cases}$

Step 5: Calculate average temperature μ_β^i and μ_γ^i of regions I_β^i and I_γ^i respectively

Step 6: Update the temperature threshold, $T_\cap^{i+1} = \frac{(\mu_\beta^i + \mu_\gamma^i)}{2} + \frac{\sigma^2}{(\mu_\beta^i - \mu_\gamma^i)} + ln\left(\frac{P_\gamma}{P_\beta}\right)$

Step 7: **if** $|T_\cap^{i+1} - T_\cap^i| < T_{ol}$; **Terminate**

Step 8: **end for**

Step 9: **return** temperature threshold $T_\cap^{i+1}$

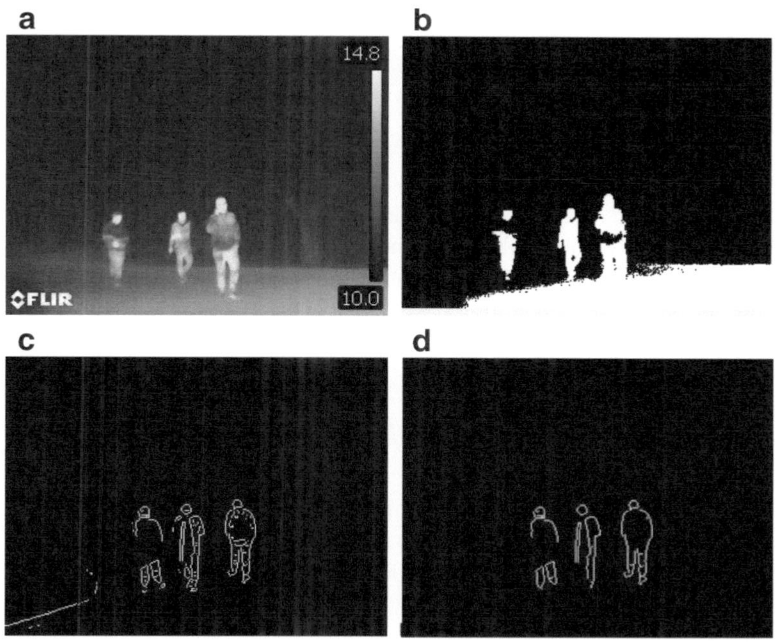

Fig. 11.2 Illustrates the various steps of proposed methodology (**a**) Thermogram, (**b**) Binary image obtained after optimal thresholding, (**c**) Edge detection applied on binary image, and (**d**) morphological opening is performed after edge detection

As shown in the algorithm, the probability of background P_β and foreground P_γ were taken as 0.4 and 0.6, respectively. To determine the optimal threshold, the value of temperature threshold $T^i_\cap$ is updated with $T^{i+1}_\cap$ at every iteration till the termination criterion $\left|T^{i+1}_\cap - T^i_\cap\right| < T_{ol}$ is satisfied. Subsequently, the temperature threshold $T^{i+1}_\cap$ is used to convert the thermogram I into a binary image, where the foreground segment (white) represents the region of interest.

11.4.3 Morphological Operations

Morphology is an approach to process and analyze the geometrical structures based on the topology, random functions, and set theory. It is mostly applicable to digital images, but can also be applicable on graphs, solids, and any other three-dimensional structures. In image processing, morphology is used to study the interaction between an image and a structuring element by using erosion and dilation as basic operations. The images are transformed using certain algorithms that improve the appearance of image. The operations like erosion, dilation, opening, and closing can help in the extraction of various features of image which is helpful in the image segmentation, noise removal, etc. In this study, the morphological opening has been performed after applying the edge detection on the binary images which gives proper geometrical information of objects. The opening of set A by structuring element B, denoted by $A \circ B$, is defined as

$$\mathrm{A} \circ \mathrm{B} = (\mathrm{A} \Theta \mathrm{B}) \bigoplus \mathrm{B} \tag{11.1}$$

where $\bigoplus$ and Θ denote the dilation and erosion, respectively. Thus, the opening is the erosion of A by B, then dilation of the result by B. As shown in Fig. 11.2, the morphological opening provides the clear geometry of objects in binary images by removing the small undesired components.

11.4.4 Pseudo-Coloring Algorithm

Although, the images acquired were the thermograms with RGB color maps, but the processing is done on gray level image to reduce the complexity of the computation. This is due to the reason that the image processing on colored images requires more computational time and increases the algorithm complexity. The gray images have poor color contrast, therefore, after the detection process has finished, the gray level images have been converted back into RGB space. In this study, the pseudo-coloring algorithm proposed by Singh and Arora has been applied (Singh and Arora 2017). In Fig. 11.3, the outcomes of different steps of proposed methodology have been

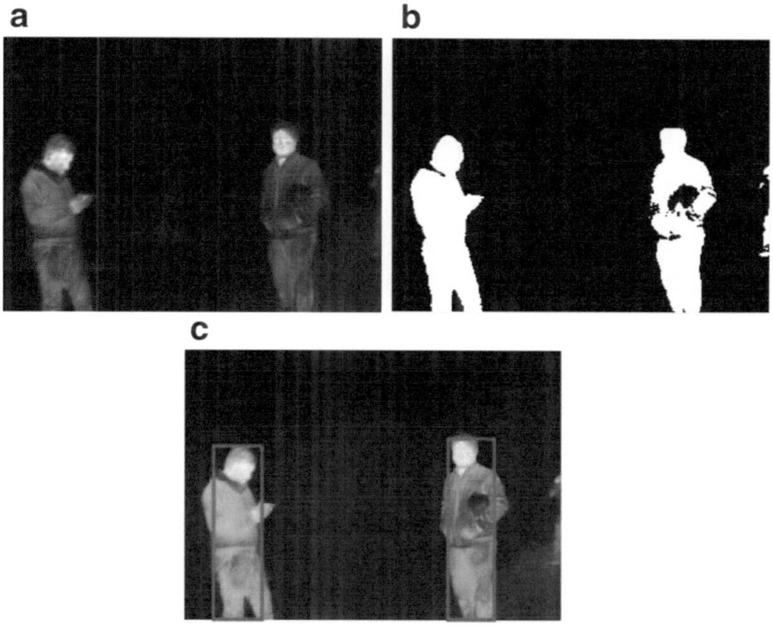

Fig. 11.3 Illustrates the outcomes of various steps of proposed methodology (**a**) Gray level image, (**b**) Binary image using optimal thresholding, and (**c**) Automatic detection and pseudo-coloring

illustrated. These algorithms are applied in the best way to get better visualization results hence enhancing the images.

11.5 Results and Discussion

This section depicts the results of the proposed approach for the detection of objects at various environmental conditions such as daytime, complete darkness, and foggy. The methodology proposed in this study has been implemented on the created dataset using MATLAB R2016. It is observed that the optimal thresholding provides satisfactory results in this environment. It removes all the irrelative background pixels and provides clear segmented view of the objects in the foreground. The visualization of images has been further improved by pseudo-coloring algorithm. The objects detected appear very bright and clear in images visually as shown in Table 11.1.

Firstly, the proposed approach has been tested on the thermograms taken in daylight condition with the objects being at different distances. Similarly, the approach has been applied on the thermograms taken in total darkness and fog.

Table 11.1 Detection of objects in different environmental conditions

Sr. No.	Visual Image	Acquired Thermograms	Optimal Threshold Image	Enhanced Image with Detected objects
1.				
2.				
3.				

Table 11.2 The outcomes of object detection in various environmental conditions

Sl. no.	Environmental conditions	Number of inputs	Correct outputs	Success
1.	Daylight	17	12	65%
2.	Night (complete darkness and fog)	36	32	88%
Overall success rate				83%

In Table 11.2, it is observed that the proposed approach shows better accuracy in thermograms of complete darkness when compared with that of daylight. During daylight, the temperature of objects of interest is usually lower or equal to the background temperature and IR reflection may be the reason for poor accuracy. Meanwhile, in case of complete darkness and fog, there is a large temperature difference between the background objects which enhances the automatic segmentation. Consequently, the approach shows the 65% and 88% accuracy during the daylight and night, respectively.

Furthermore, the performance of the proposed methodology has been evaluated by comparing the outcomes with that of other standard approaches. As a result, the accuracies obtained are 71%, 53% and 83% by applying the edge detection, Otsu's technique, and proposed approach, respectively. In Table 11.3, it is observed that the edge detection method detects more objects than that actually exists, whereas Otsu's technique provides poor detection and vague geometrical information of objects. Meanwhile, the proposed approach detects the objects of interest accurately with clear boundary information.

Table 11.3 Comparison of outcomes of proposed method with other methods

Sr. No.	Edge Detection	Otsu Thresholding	Optimal thresholding (Proposed Method)
1.	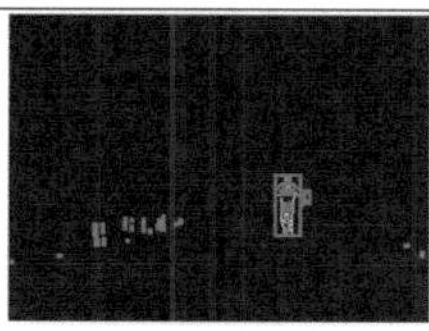		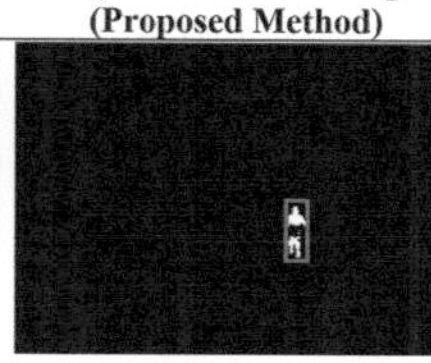
2.	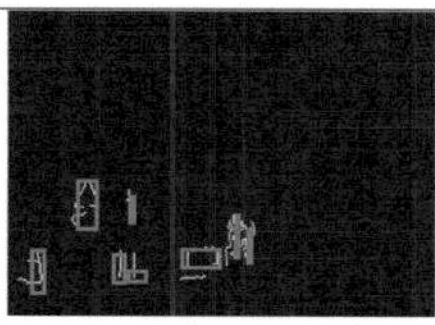		
3.	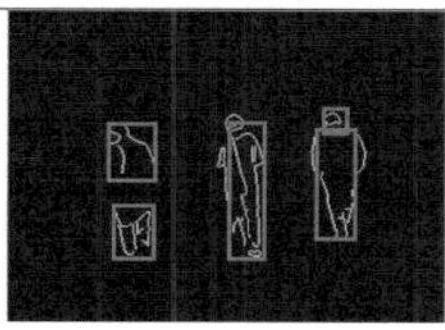		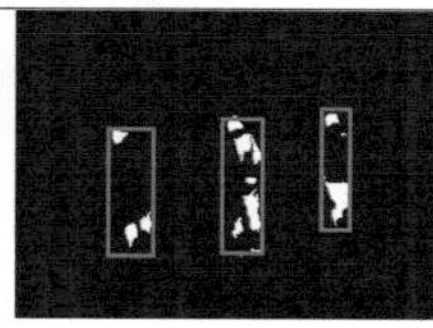

11.6 Conclusion

The results obtained from the proposed methodology shows that it is successful in detecting the objects in poor environmental conditions. The overall accuracy of proposed method is 83%, where it shows good detection in the dataset of complete darkness and fog when compared with the dataset of daylight. The proposed algorithm is very simple and effective. In addition, it requires very less storage space and computation time. The effectiveness of proposed approach has been analyzed by comparing the outcomes with that of standard techniques. Consequently, the proposed approach shows better results. Also, the issues and challenges faced by thermal based UAVs are discussed. Moreover, the proposed algorithm can efficiently detect the recent accidental vehicles due to rise in temperature which further helps to give automatic call to rescue team. Hence, the thermal imaging can play a significant role in UAVs for efficient road patrolling.

References

Arora S, Pandey R (2016) Applications of morphological operators using image morphological algorithms. SSRG Int J Electron Commun Eng 3:107–110

Austin R (2011) Unmanned aircraft systems: UAVS design, development and deployment, vol 54. Wiley, Somerset

Berni JA, Zarco-Tejada PJ, Suárez Barranco MD, Fereres Castiel E (2009) Thermal and narrow-band multispectral remote sensing for vegetation monitoring from an unmanned aerial vehicle. Institute of Electrical and Electronics Engineers

Bertozzi M, Broggi A, Fascioli A, Graf T, Meinecke MM (2004) Pedestrian detection for driver assistance using multi resolution infrared vision. IEEE Trans Veh Technol 53:1666–1678. conservation. Sensors 16: 97

Das A, Ghasemzadeh A, Ahmed MM (2019) Analyzing the effect of fog weather conditions on driver lane-keeping performance using the SHRP2 naturalistic driving study data. J Saf Res 68:71–80

Everaerts J (2008) The use of unmanned aerial vehicles (UAVs) for remote sensing and mapping. Int Arch Photogramm Remote Sens Spat Inf Sci 37:1187–1192

Gade R, Moeslund TB (2014) Thermal cameras and applications: a survey. Mach Vis Appl 25 (1):245–262

Gonzalez LF, Montes GA, Puig E, Johnson S, Mengersen K, Gaston KJ (2016) Unmanned Aerial Vehicles (UAVs) and artificial intelligence revolutionizing wildlife monitoring and conservation. Sensors 16:97

Goubet E, Katz J, Porikli F (2006) Pedestrian tracking using thermal infrared imaging. In: Infrared technology and applications XXXII 6206: 62062C. International Society for Optics and Photonics

Jayadevan R, Navas KA (2014) Automated pseudo-coloring of grayscale images based on contour let transform. In: Communication: Signal Processing and Networking (NCCSN), IEEE, pp 1–6

Luo Y, Remillard J, Hoetzer D (2010) Pedestrian detection in near-infrared night vision system. In: Intelligent vehicles symposium IV, 2010 IEEE, pp 51–58

Nishar A, Richards S, Breen D, Robertson J, Breen B (2016) Thermal infrared imaging of geothermal environments and by an unmanned aerial vehicle (UAV): a case study of the Wairakei–Tauhara geothermal field, Taupo, New Zealand. Renew Energy 86:1256–1264

Omar M, Zhou Y (2007) Pedestrian tracking routine for passive automotive night vision systems. Sens Rev 27:310–316

Pandey RK, Mathurkar SS (2017) Implementation of parallel morphological filter with different structuring elements. In: 2017 IEEE International Conference on Power, Control, Signals and Instrumentation Engineering (ICPCSI), IEEE, pp 1450–1455

Piniarski K, Pawłowski P, Dąbrowski A (2014) Pedestrian detection by video processing in automotive night vision system. In: Signal processing: algorithms, architectures, arrangements, and applications, IEEE, pp 104–109

Quater PB, Grimaccia F, Leva S, Mussetta M, Aghaei M (2014) Light unmanned aerial vehicles (UAVs) for cooperative inspection of PV plants. IEEE J Photovolt 4:1107–1113

Rajkumar S, Mouli PC (2015) Pedestrian detection in infrared images using local thresholding. In: Electronics and communication Systems (ICECS), 2015 2nd international conference, IEEE pp 259–263

Riggan PJ, Hoffman JW (2003) FireMapper™: a thermal-imaging radiometer for wildfire research and operations. In: Proceedings of the IEEE aerospace conference

Road Crash Statistics (2016). Available at http://savelifefoundation.org/wp-content/uploads/2017/09/Road-Crash-Factsheet_SLF_2016.pdf

Saito H, Hagihara T, Hatanaka K, Sawat T (2008) Development of pedestrian detection system using far- infrared ray camera. SEI Tech Rev Engl Ed 66:112

Singh SK (2017) Road traffic accidents in India: issues and challenges. Transportation Res Procedia 25:4708–4719

Singh J, Arora AS (2017) Contrast enhancement algorithm for IR Thermograms using optimal temperature Thresholding and contrast stretching. In Advances in machine learning and data science, Springer, pp 361–368

Singh J, Arora AS (2018) An automated approach to enhance the thermographic evaluation on orofacial regions in lateral facial thermograms. J Therm Biol 71:91–98

Tsuji T, Hattori H, Watanabe M, Nagaoka N (2002) Development of night-vision system. IEEE Trans Intell Transp Syst 3:203–209

Zhang J, Jung J, Sohn G, Cohen M (2015) Thermal infrared inspection of roof insulation using unmanned aerial vehicles. Int Arch Photogramm Remote Sens Spat Inf Sci:40, 381–386

Chapter 12
Fusion and Enhancement Techniques for Processing of Multispectral Images

Ashwani Kumar Aggarwal

Abstract With recent advances in sensor technology, several sensors are being extensively used to continuously monitor agriculture areas to have better yield. These sensors, which include Red, Green, Blue(RGB) cameras, thermal cameras, infrared radiation (IR) cameras, multispectral sensors, and hyperspectral sensors are either ground based sensors or unmanned aerial vehicle (UAV) based sensors. While ground based sensors do not have limitation of sensor size and weight in most of the cases, however, the field of data capture of such sensors is limited as compared to sensors mounted on UAVs. In this chapter, various techniques to capture imagery of agriculture fields using several sensors are discussed with special focus on enhancement and fusion techniques used for processing of multispectral images. Although several methods are available in literature, which work in either spatial domain or frequency domain to enhance multispectral images, however, each of those methods suffer from their own drawbacks. Existing multispectral image fusion methods directly take images captured using different wavelengths of electromagnetic spectrum and fuse them based on template matching, feature based or several other techniques. Such methods are quick in performing fusion task, however, the required information is sometimes missed or redundant information is embedded in the fused image causing large size of multispectral image. The images are enhanced based on histogram equalization and homomorphic filtering before applying fusion algorithm. Experiments were conducted on a range of multispectral images and fusion results obtained are promising.

Keywords Multispectral images · Homomorphic filtering · Fusion · Histogram equalization · Remote sensing

A. K. Aggarwal (✉)
Sant Longowal Institute of Engineering and Technology (SLIET),
Longowal, Sangrur, Punjab, India

R. Avtar, T. Watanabe (eds.), *Unmanned Aerial Vehicle: Applications in Agriculture and Environment*, https://doi.org/10.1007/978-3-030-27157-2_12

12.1 Introduction

Multispectral image analysis is used to depict changes in environment, food quality assessment, weather forecasting, and provide local as well as global solutions to various issues related to effects of such factors. Multispectral images are widely used in remote sensing applications such as for extraction of leaf area, estimation of solar panel coverage, urban population, etc. The multispectral images because of their high spectral resolution find their applications in agriculture for crop quality assessment. As a single band image cannot capture all the useful information needed for remote sensing applications, several wavelengths of the electromagnetic spectrum ranging from near infrared, mid infrared, far infrared, visible region, and others are used to obtain images in different bands. Each of the band gives information out of which some information is common with other bands whereas part of the information belongs to its particular band. Multispectral images have advantage over RGB images that their spectral resolution is higher and hence these images contain much more information than that contained in RGB images. Because of several bands of multispectral images, the image size of multispectral image is also higher as compared to image size of RGB images. Each band of image is equivalent to intensity image having a matrix of pixel values, which represent the inherent properties of the scene captured. The pixel values are dependent on the amount of radiations absorbed by the portion of scene, which is mapped to that pixel in multispectral image. A sample multispectral image of 30 m resolution is shown in Fig. 12.1.

Researchers working in span of remote sensing applications choose a subset of such bands to meet their requirement. As each band of the multispectral image represents information based on interaction of scene with radiation in that particular wavelength band, its processing needs different algorithms suitable for enhancement

Fig. 12.1 Multispectral image of resolution 30 m

$$LANDSAT = \begin{cases} Band\ 1 & 0.43 & - 0.45\ \mu m, \\ Band\ 2 & 0.45 & - 0.51\ \mu m, \\ Band\ 3 & 0.52 & - 0.60\ \mu m, \\ Band\ 4 & 0.63 & - 0.68\ \mu m, \\ Band\ 5 & 0.84 & - 0.88\ \mu m, \\ Band\ 6 & 1.56 & - 1.66\ \mu m, \\ Band\ 7 & 2.10 & - 2.30\ \mu m \end{cases}$$

Fig. 12.2 Typical values for wavelength in seven bands of LANDSAT data

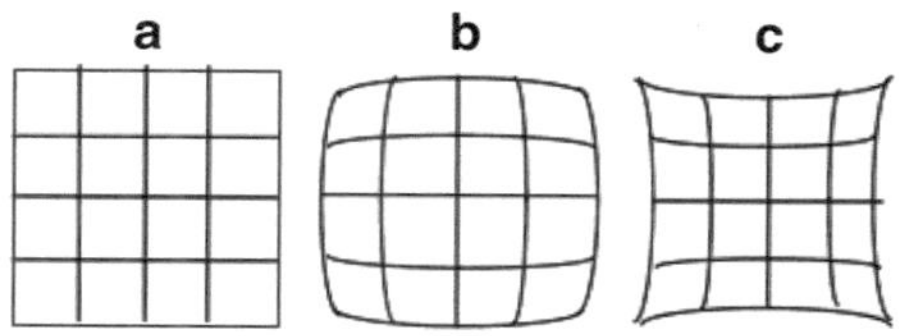

Fig. 12.3 (**a**) Distortion-free image (**b**) Barrel distortion (**c**) Pincushion distortion

of image. LANDSAT dataset contains several bands in multispectral images. Typical values for wavelength in seven bands of LANDSAT data are shown in Fig. 12.2.

Multispectral image data needs enhancement because it suffers from many geometric and radiometric distortions at the time of capture. Several types of distortion in multispectral image data are shown in Fig. 12.3. Such distortions occur mainly due to sensor noise, atmospheric disturbances, and non-uniform solar illumination among many others. Therefore, raw data cannot be used as such for classification of multispectral images and if used as such, would lead to false image analysis results. The role of image enhancement in multispectral images is to apply a combination of techniques so that processed image is free from such distortions. The sources of geometric distortion, which occurs due to sensor characteristics, are aspect ratio, misalignment in detector geometry and non-linearity in the camera modules.

Several commercial and open source softwares are available to processes multispectral image data. Out of a list of such softwares, a few of the widely used softwares are listed below.

Sentinel Toolbox Sentinel toolbox consists of three sub-categories. Sentinel-1 toolbox is used to process synthetic aperture radar (SAR) data while Sentinel-2 and Sentinel-3 toolboxes are used to process multispectral data. Processing of data includes filtering, registration and other morphological operations on the images.

SAGA GIS – System for Automated Geoscientific Analyses SAGA GIS have vast library of functions for classification of multispectral images which include classification based on supervised learning methods, viz. support vector machines, neural networks, decision trees, etc.

Opticks Opticks software supports SAR, hyperspectral and multispectral and other types of remote sensing data. It is used for image enhancement using edge operators, histogram equalization, etc. The software also is a good tool for traditional image classification based on unsupervised, supervised classifiers as well as based on feature-based classifiers.

gvSIG gvSIG software includes a tool known as Tasseled Cap for analyzing vegetation growth and its health, which is quite useful in agriculture field. The vegetation index calculation from multispectral images using gvSIG helps a researcher to analyze the health of leaves of the plants.

12.2 Related Work

A pan sharpening Amélioration de la résolution spatiale par injection de structures (ARSIS) technique is used for fusion of multispectral images in (Sylla et al. 2014). The authors apply their algorithm on Medium Resolution Imaging Spectrometer (MERIS) dataset and compare their results with other methods of fusion. They used synthetic as well as real datasets on their algorithm. The authors found that their method is more suitable for water region than that for land region. A multiagent approach is used for fusion of satellite images in (Farah et al. 2008). The authors use different abstraction levels to achieve fusion task. The authors show an improvement in classification of satellite images using their method. A Bayesian framework is used for fusion of multispectral and panchromatic images in (Khademi and Ghassemian 2017).

The fusion task is formulated as a constrained optimization problem, which is solved using complex conjugate optimization method. The method is compared with other pan sharpening methods. Fusion of multispectral images using pixel manipulation including averaging, weighted least squares, neighborhood filtering, etc. is evaluated in (Hryvachevskyi et al. 2018). The authors compare nine different methods based on several performance indices for fusion of multispectral images. A dictionary pair localization is used for fusion of multispectral images in (Liang et al. 2017). The method has the advantage of simplicity and lower computation cost. Up-sampling and spatial degradation of the two images are done for the fusion task. Spectral angle error and Peak signal-to-noise-ratio (PSNR) are used as performance indices for comparison of their method with other fusion methods. To take advantage of both spatial resolution from multispectral images and spectral resolution from hyperspectral images, an approach for fusion of hyperspectral and multispectral images is given in (Palsson et al. 2017). Spectral image is split into many sub-images using low rank method. An update algorithm based on abundance and end-members is used to fuse the images. The fusion algorithm is tested on three different datasets based on different evaluation parameters. A convolutional network is used for fusion of multispectral and hyperspectral images in (Wei et al. 2014). To reduce the computation time for the fusion to take place, dimensionality reduction of hyperspectral image is done. The method is also robust to additive noise. The authors also performed a comparative analysis of their methods with other methods in case of hyperspectral image corrupted with noise. A sparse representation is used for fusion of hyperspectral and multispectral images in (Qian et al. 2017).

An inverse problem is formulated to obtain fused image from hyperspectral and multispectral images. The method uses subspace learning and dictionary learning.

The results are compared with other methods for image fusion using simulated data. Morphological operators have been used in (Restaino et al. 2016) to fuse multispectral and panchromatic images. The authors use four different datasets to test their method. Reduced resolution and full resolution of images are used for the comparative analysis. A global regression based fusion algorithm is proposed in (Liu et al. 2018). The authors use red, blue, green color channels and near infrared images along with panchromatic images for the fusion task. Several multispectral images are processed and the method is optimal in minimal mean square sense. Linear transformation is used for multispectral image resolution in (Inglada 2016). The authors use canonical variable substitution method for enhancement of multispectral images. Different weights are assigned in the linear transformation based on the scene properties. The authors compare their results with intensity hue saturation method for enhancement of multispectral images. Offset sparsity decomposition based image enhancement for multispectral images is done in (Tian et al. 2016). The method splits the multispectral image into three constituent images, viz. enhanced image, offset image, and error image. The method works by estimating enhanced image from given image as observed image. Gaussian noise is added to estimate the robustness of the method. An enhancement technique based on unmixing for hyperspectral images is proposed in (Lu and Weng 2004). The method is tested on real data for enhancement evaluation. A hybrid approach based on Fuzzy statistics variables and Principal Component Analysis (PCA) is used for enhancement of multispectral images in (Lu and Weng 2004).

The outlier problem is overcome using this method (Sun et al. 2018). The authors compare their method with classical PCA technique on LANDSAT data. Image enhancement based on neural networks and histogram equalization is done in (Doering 2016). The method is tested on synthetic data as well as on real data. Contrast enhancement for multispectral images is achieved in (Cheng et al. 2016). An operating point is chosen in RGB color space of the multispectral image and histogram modification is done to achieve contrast enhancement. Weighted least squares method is used for enhancement of multispectral images in (Zhou et al. 2016). The method is applied on satellite images as well as on spatially degraded images. Sparse unmixing is done on hyperspectral and multispectral images for image enhancement in (Yoshioka et al. 2015). LANDSAT images and Sentinel images are fused for continuous monitoring in (Xiao et al. 2018). A Laplacian pyramid is used to merge multispectral images with panchromatic images in (Corner et al. 2003).

12.3 Image Enhancement and Fusion Techniques

Multispectral Image enhancement is achieved either in the spatial domain or in frequency domain. Hence enhancement methods for multispectral images can be divided into two categories.

- Spatial domain enhancement
- Frequency domain enhancement

Spatial domain enhancement methods work directly on pixel data. Each pixel of enhanced image is obtained either from a single pixel of input image or from a combination of pixels in some neighborhood $M \times M$ of a pixel. Accordingly, the enhancement is known as point processing and neighborhood processing. The spatial domain methods are represented as follows:

$$g(x, y) = T\,[f(x, y)] \tag{12.1}$$

where $f(x, y)$ and $g(x, y)$ in Eq. (12.1) are input image and enhanced image respectively. T [.] is the transformation matrix, which maps input image to output image. Piecewise linear transformations are sometimes used instead of a single transformation applied to whole image. The whole image is sub-divided into many regions of interest and for each region of interest a separate linear transformation is chosen to enhance the image. Several types of spatial domain filters are used to enhance contrast, increase brightness and to remove blurring effects in input images. Some of the methods use histogram equalization based image enhancement (Liu and Eom 2013). The histogram is skewed in the lower side of grayscale values and stretched in the higher side of grayscale values. Fig. 12.4 shows histogram of Band 1 of a multispectral image, which is used for enhancement task.

Frequency domain enhancement methods work by first converting spatial domain images into frequency domain, applying suitable filtering techniques and again converting them back to spatial domain. The advantage of conversion from spatial domain to frequency domain is reducing the computation time involved in the process of enhancement as convolution in spatial domain takes lot of computation whereas Fast Fourier Transform (FFT) and frequency domain methods take lesser computation time. Low pass frequency domain filtering involves removal of

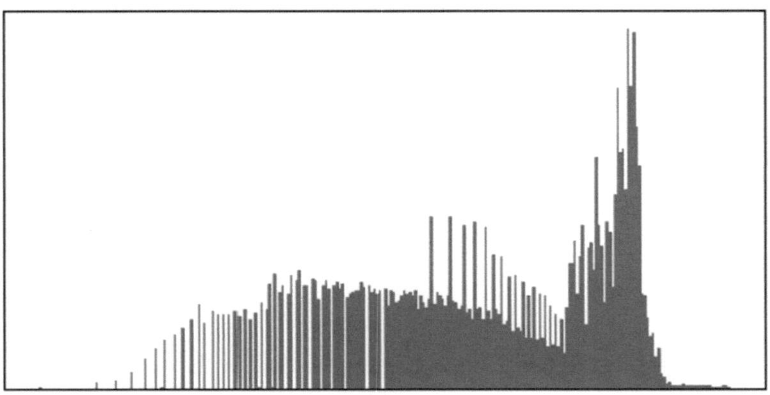

Fig. 12.4 Histogram of Band 1 of multispectral image

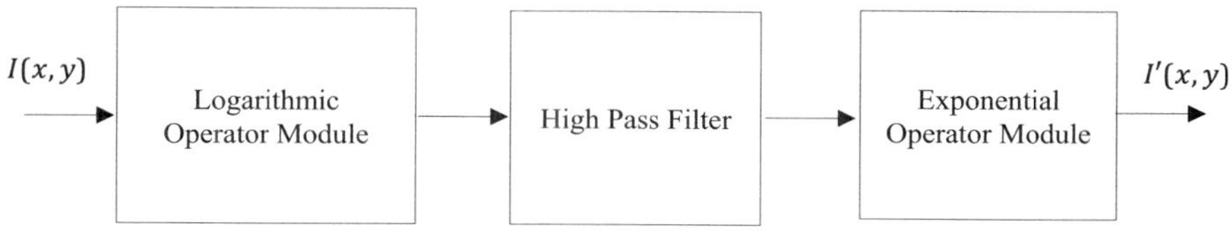

Fig. 12.5 Flowchart of enhancement technique for distorted multispectral image

high frequency components in image. It causes blurring of image as sharp transitions, which account to high frequency data are removed by low pass filters. Therefore, appropriate filter design parameters need to be chosen for enhancement of images (Huang et al. 2018).

Homomorphic filtering is applied on the multispectral image, which works in logarithmic domain. The input image is given to logarithmic operator module and it is filtered with high pass filter (Wei et al. 2014). The filtered image is then given to exponential operator module to get an image that is filtered image of original multispectral image. Using homomorphic filtering the non-uniform illumination effect is cancelled and the image is enhanced.

High pass filter used on homomorphic filter is in frequency domain which operates on logarithmic input image and gives an image as logarithm of output image. Flowchart of homomorphic filtering is shown in Fig. 12.5.

Image fusion is a task of combining all the information obtained from several images into a single image. Some of the information contained in different images is complementary whereas some information in images is common (Su et al. 2018). A good fusion method removes any redundant information in the fused image while retaining all the complementary information in the fused image. Image fusion is primarily of following types:

- Multi-view Image Fusion
- Multi-sensor Image Fusion
- Multi-focus Image Fusion
- Multi-temporal Image Fusion

Multi-view Image Fusion This type of image fusion is required when data is collected by a single sensor from different viewpoints. The sensor is displaced in horizontal and vertical directions to obtain several images. These images are then combined so that overlapping image regions are used to fuse the constituent images. More number of images increases complexity but at the same time it improves fusion results.

Multi-sensor Image Fusion Several sensors operating on different wavelength regions of electromagnetic spectrum are used to capture a scene. As different sensors are not aligned properly with each other, there is sensor misalignment, which leads to a need for fusion of multispectral images (Huang et al. 2015).

Multi-focus Image Fusion In order to cover various regions in the scene, sensor focus is changed and many images are collected. These images need to be fused into single image to obtain information of the different regions in the scene for better analysis.

Multi-temporal Image Fusion To capture the changes in the scene with time, sensor data is collected at different times using some sampling frequency. The images obtained during a span of time are then fused into a single image for further analysis (Hupple et al. 2018).

Multispectral image fusion is based on following common methods. Each of the methods has its own benefits and drawbacks.

- Averaging method
- Principal Component Analysis (PCA) based method
- Discrete Wavelet Transform (DWT) based method

Averaging Method Averaging method works by taking pixel-wise average of intensities in corresponding images of candidate images.

$$F(x,y) = \frac{1}{\mathrm{N}} \sum_{i=1}^{N} A_i(x,y) \tag{12.2}$$

where $A_i(x, y)$ in Eq. 12.2 is one of the constituent images used in the fusion process and $F(x, y)$ is the fused image. The dimensions of two images to be fused need to be same for the fusion algorithm to work. In case of different image sizes, the resizing of images are carried out before fusion process. This type of fusion method is not suitable when any of the images suffers from multiplicative noise (Saura et al. 2019). Averaging method takes into account all the pixels in the two images which are to be fused and corresponding pixels are used in the computation to obtain the pixel value in the output image.

Principal Component Analysis (PCA) Based Method PCA is a statistical method, which uses orthogonal transformation. It converts a set of data, which is correlated into a set of uncorrelated variables called principal components. This method works by converting a higher dimensional data of dimensions M into lower dimensional data of dimensions N, and evaluating the error term as given in Eq. (12.3):

$$\in^2 = \frac{1}{M} \sum_{i=1}^{M} (X_i)^2 - \sum_{j=1}^{N} b_j^T \left(\frac{1}{M} \sum_{i=1}^{M} X_i X_i^T \right) b_j \tag{12.3}$$

PCA based image fusion methods suffer from drawback that image data should be linearly correlated for the fusion algorithm to work. Another drawback of PCA based multispectral image fusion method is that it is scale variant. Therefore, it does not change the scale of the data and rotates the data with a transformation such that transformed data is along orthogonal vectors, which are uncorrelated vectors (Saritha and Kumar 2015). The number of principal components are chosen in such a way that useful information in the output image is captured while achieving the dimensionality reduction using PCA approach (Herrero et al. 2014).

Discrete Wavelet Transform (DWT) Based Method The wavelet transform works in multiscale domain. DWT first blurs the image in horizontal direction and then down-samples each row of the image by an integer number using filters. This is one level of decomposition and then process is repeated for several levels. It decomposes the image into low-low (LL), low-high (LH), high-low (HL) and high-high (HH) frequency bands in horizontal and vertical directions.

The choice of DWT coefficients depends on the image data. The DWT coefficients are calculated on pixel level. The process is repeated in vertical direction and multiscale decomposition is done as in the horizontal direction (Qin et al. 2013). DWT based image fusion works by transformation of spatial domain image data into wavelet coefficients. A block diagram for the fusion technique based on DWT is shown below in Fig. 12.6.

Fusion of multispectral images is achieved either using time domain or frequency domain methods. As many bands need to be fused for the required information to be put into the fused image, the size of the image becomes large (Hagag et al. 2016). It is of significant importance to keep the image size to minimum but keeping all the necessary information (Hong et al. 2018).

Hybrid Method This method works by taking advantage of averaging method as well as dimensionality reduction using Principal Component Analysis. The averaging method considers pixels of constituent images which are to be fused into a single

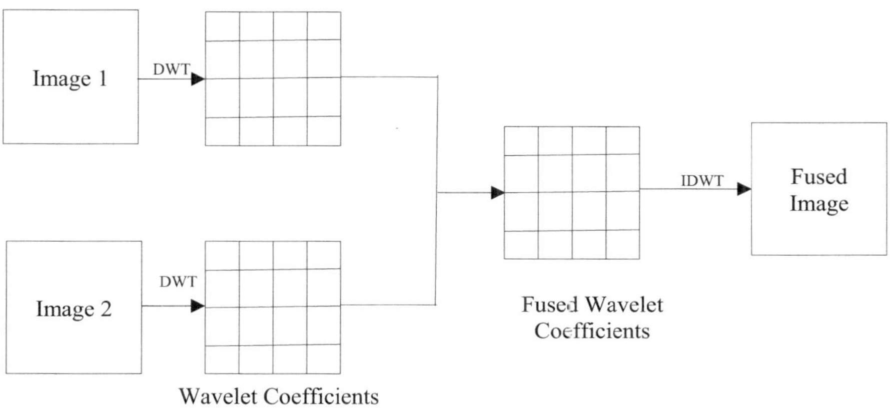

Fig. 12.6 DWT based image fusion

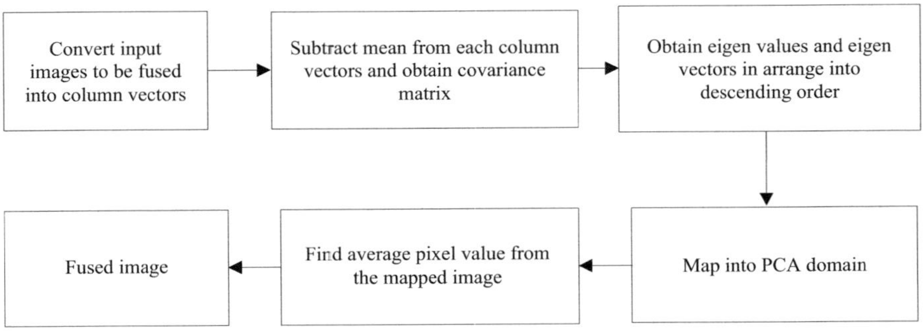

Fig. 12.7 Hybrid method to image fusion

image whereas, PCA based multispectral image fusion works by transforming the image data around vectors which are orthogonal and are uncorrelated. A weighted average method is sometimes used, where different weights are assigned to each of the fusion method. A proper choice of weights for each of the fusion method is performed using optimization technique. The optimization criterion is chosen depending on parameters of the fused image. A block diagram for the hybrid approach to image fusion is shown in Fig. 12.7. The covariance matrix for each of the constituent images is obtained and then average pixel values from PCA mapped image are used in the fusion process.

The hybrid method is implemented to multispectral image fusion using the following steps:

Step 1. Finding the Reference image. A reference image of the size of input image is found and used in the fusion process.

Step 2. Select a grid fitting the reference image. The grid should have the same dimensions as that of the reference image.

Step 3. Choose control points in the given image. The control points are chosen in such a way that these control points do not lie on a straight line.

Step 4. Apply transformation to given image. Using an initial set of transformation parameters, given image is transformed into new image.

Step 5. Resample given image and assign new grayscale pixel values if termination criterion is met. Else, update the transformation matrix and calculate average minimum distance between two images.

Step 6. After the set termination criterion is met, the output image is the fused image obtained from its constituent images.

The hybrid method of image fusion considers advantages of many fusion methods to obtain a fused image.

A flowchart for the image enhancement for distorted image is shown in Fig. 12.8.

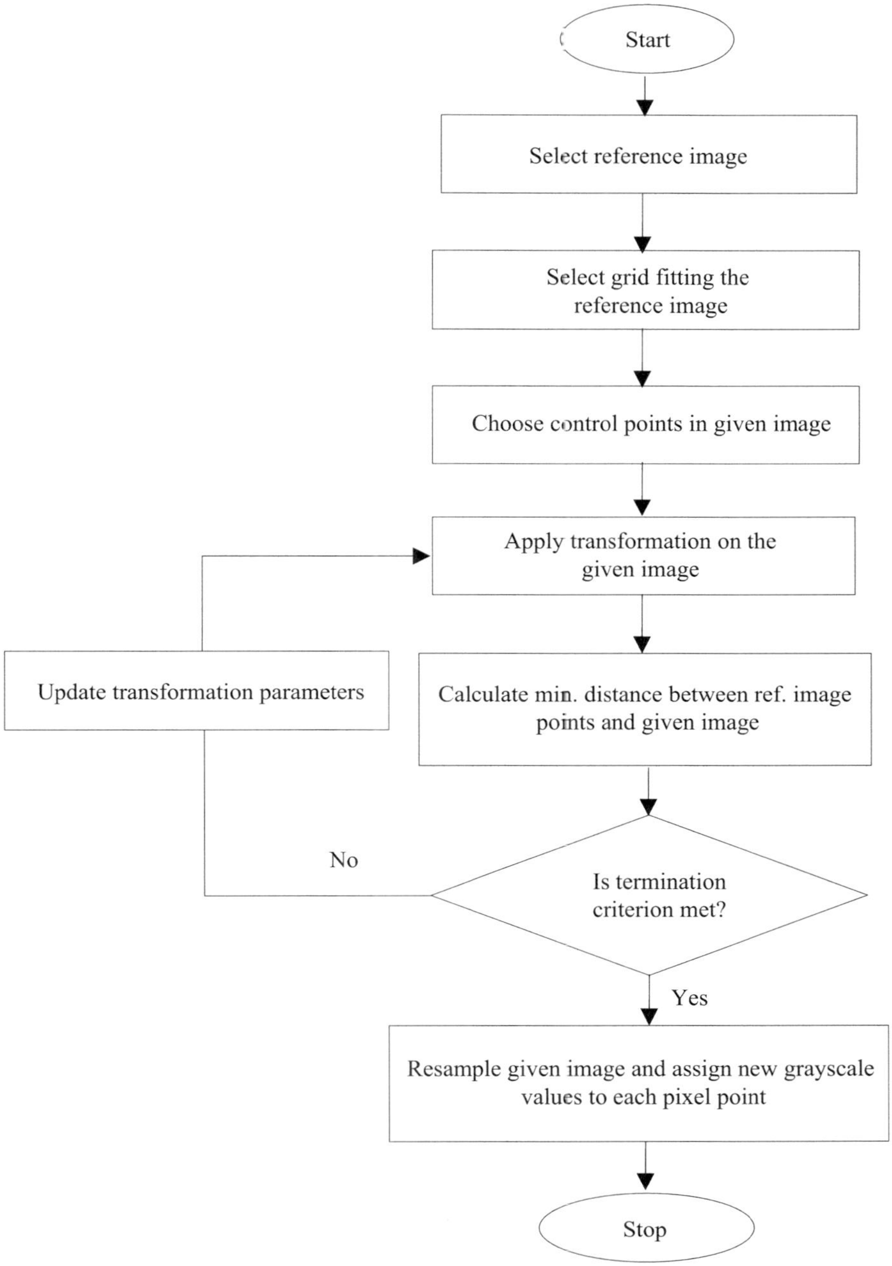

Fig. 12.8 Flowchart of enhancement technique for distorted multispectral image

12.4 Performance Measures

Mean Square Error (MSE) Mean Square error is the average of the squared errors between actual and calculated data. MSE calculation is a common metric to estimate the performance of image enhancement techniques. MSE is given by Eq. (12.4) below:

$$\mathrm{MSE} = \frac{1}{mn} + \sum_{i=0}^{m-1} \sum_{j=0}^{n-1} [I(i,j) - K(i,j)]^2 \tag{12.4}$$

where $I(i,j)$ and $K(i,j)$ are reference image and fused image, respectively. M and N are number of rows and number of columns in the input images. PSNR can be obtained from MSE using $\mathrm{PSNR} = 10\log\left(\frac{R^2}{\mathrm{MSE}}\right)$. Where R is maximum pixel value in the image data. This performance measure is used to compare the performance of many algorithms.

Structural Similarity Index (SSIM) SSIM is structural similarity index and is a perceptual metric. SSIM is a metric to quantify the performance of fusion method. SSIM is given by Eq. (12.5) below:

$$\mathrm{SSIM}(x,y) = \frac{\left(2\mu_x\mu_y + c_1\right)\left(2\sigma_{xy} + c_2\right)}{\left(\mu_x^2 + \mu_y^2 + 1\right)\left(\sigma_x^2 + \sigma_y^2 + +c_2\right)} \tag{12.5}$$

Moments Mathematically moments can be defined as:

$$E_{ri} = \frac{1}{N}\sum_{i=1}^{N} I_{ij} \tag{12.6}$$

$$\sigma_{r,i} = \sqrt{\frac{1}{N}\sum_{\mathrm{j}=1}^{N}\left(I_{ij} - E_{ri}\right)^2} \tag{12.7}$$

$$S_{r,i} = \sqrt[3]{\frac{1}{N}\sum_{\mathrm{j}=1}^{N}\left(I_{ij} - E_{ri}\right)^3} \tag{12.8}$$

where I_{ij} and E_{ri} in Eqs. (12.6, 12.7, and 12.8) are reference image and fused image, respectively. N is the number of pixels in the image. Chebyshev moments, geometric moments, rotational moments, and complex moments are also used apart from above-mentioned moments as performance measures.

Entropy The concept of information theory is used to evaluate the performance measure of fusion method. The entropy of the image changes with change in texture of the image. A low entropy means more homogeneity in the image. A high entropy image means more information content in the image. The entropy or the average information of an image measures the randomness in the image (Kumar et al. 2013). Each of the input images and fused image are considered as random variables with discrete probability distribution (Martins et al. 2015). The entropy $H(x)$ of a random variable x is defined by:

$$H(x) = \sum\nolimits_x p(x) \log p(x) \tag{12.9}$$

where $p(x)$ in Eq. (12.9) is probability density function of random variable x. The entropy is one of the common metrics used to evaluate the performance of image fusion methods.

12.5 Results and Discussion

The multispectral image is enhanced using homomorphic filter, which works in logarithmic domain. The resultant filtered image is shown in Fig. 12.9. The multiplicative noise is removed from the original multispectral image and image quality is improved for remote sensing purposes. The output image is blurred to some extent but the multiplicative noise is removed from the input image. The blurring of the image is removed with any of the deblurring methods used in multispectral image processing.

Fig. 12.9 Homomorphic filtered image

The enhanced image obtained after homomorphic filtering is segmented using Otsu's thresholding method. The resultant image obtained after segmentation is shown below in Fig. 12.10.

The enhanced image is taken as one of the constituent images in the fusion process and fused image is obtained which preserves the information of each of the constituent images while removing the redundant is also segmented using hybrid thresholding method and is shown below in Fig. 12.11.

The segmented image is taken as one of the constituent images in the fusion process and fused image is obtained which preserves the information of each of the constituent images while removing the redundant information.

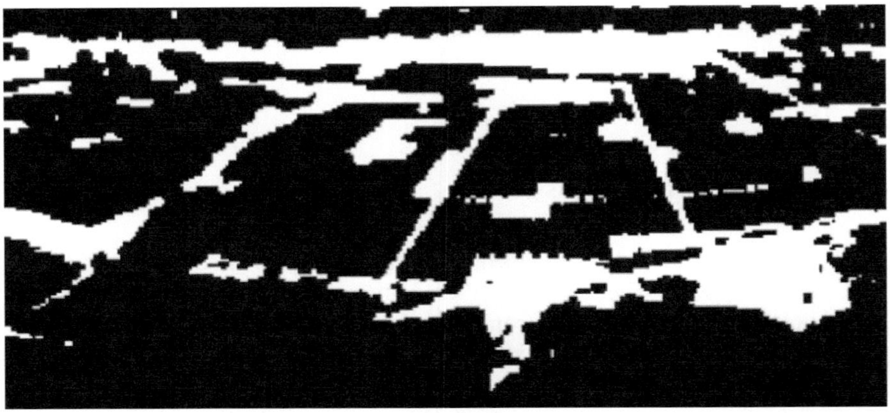

Fig. 12.10 Segmented image with Otsu's thresholding

Fig. 12.11 Segmented image with hybrid method

12.6 Conclusion

Multispectral imagery is useful for finding the chlorophyll content in leaves, detecting various diseases in crops, estimating growth of plants. The raw images captured from multispectral sensors suffer from many distortions and single image is sometimes not that useful for analysis. In this chapter, enhancement methods for removing several distortions in multispectral images are discussed. In addition, different fusion techniques used for fusion of multispectral images are discussed with performance metrics to quantitatively measure the performance of different fusion techniques. Fusion of multispectral images keeps the complementary information contained in its constituent images into the fused image for analysis of various aspects of crops.

References

Cheng J-H, Sun D-W, Qu J, Pu H (2016) Developing a multispectral imaging for simultaneous prediction of freshness indicators during chemical spoilage of grass carp fish fillet. J Food Eng 182:9–17. https://doi.org/10.1016/j.jfoodeng.2016.02.004

Corner BR, Narayanan RM, Reichenbach SE (2003). Noise estimation in remote sensing imagery using data masking. Int J Remote Sens 24(4):689–702. Retrieved from https://doi.org/10.1080/01431160210164271

Doering D, Vizzotto M, Bredemeier C, da Costa C, Henriques R, Pignaton E, Pereira C (2016) MDE-based development of a multispectral camera for precision agriculture. IFAC-Papers OnLine 49(30):24–29. Retrieved from http://www.sciencedirect.com/science/article/pii/S2405896316325551. (4th IFAC Symposium on Telematics Applications TA 2016). https://doi.org/10.1016/j.ifacol.2016.11.117

Farah IR, Boulila W, Ettabaa KS, Ahmed MB (2008) Multiapproach system based on fusion of multispectral images for land-cover classification. IEEE Trans Geosci Remote Sens 46 (12):4153–4161. https://doi.org/10.1109/TGRS.2008.2001554

Hagag A, Hassan E, Amin M, Abd El-Samie F, Fan X (2016) Satellite multispectral image compression based on removing sub-bands. Optik Int J Light Electron Opt 131:1023–1035. https://doi.org/10.1016/j.ijleo.2016.11.172

Herrero-Huerta M, Hernandez D, Rodrguez-Gonzlvez P, Gonzlez-Aguilera D, Gonzalez-Piqueras J (2014) Vicarious radiometric calibration of a multispectral sensor from an aerial trike applied to precision agriculture. Comput Electron Agric 108:2838. https://doi.org/10.1016/j.compag.2014.07.001

Hong K, Liu X, Liu G, Chen W (2018) Detection of physical stress using multispectral imaging. Neurocomputing 329:116–128. https://doi.org/10.1016/j.neucom.2018.10.011

Hryvachevskyi A, Prudyus I, Lazko L, Fabirovskyy S (2018) Methods of multispectral image fusion at the pixel level. In 2018 14th international conference on advanced trends in radioelecrtronics, Telecommunications and Computer Engineering (TCSET), pp 102–107. https://doi.org/10.1109/TCSET.2018.8336165

Huang W, Li J, Wang Q, Chen L (2015) Development of a multispectral imaging system for online detection of bruises on apples. J Food Eng 146:62–71. Retrieved from http://www.sciencedirect.com/science/article/pii/S0260877414003641. https://doi.org/10.1016/j.jfoodeng.2014.09.002

Huang Y, Xin Chen Z, Yu T, Zhi Huang X, Fa Gu X (2018) Agricultural remote sensing big data: Management and applications. J Integr Agric 17(9):1915–1931. Retrieved from http://www.

sciencedirect.com/science/article/pii/S2095311917618598s. https://doi.org/10.1016/S2095-3119(17)61859-8

Hupple CW, Morscher S, Burton NC, Pagel MD, McNally LR, Crdenas-Rodrguez J (2018) A light-fluence-independent method for the quantitative analysis of dynamic contrast-enhanced multispectral optoacoustic tomography (DCE MSOT). Photo-Dermatology 10:54–64. Retrieved from http://www.sciencedirect.com/science/article/pii/S2213597917300496. https://doi.org/10.1016/j.pacs.2018.04.003

Inglada J (2016) Land cover mapping from optical images. In: Land surface remote sensing in agriculture and Forest. Elsevier, pp 81–111

Khademi G, Ghassemian H (2017) Bayesian fusion of multispectral and panchromatic images. In 2017 10th Iranian conference on machine vision and image processing (MVIP), pp 20–25. https://doi.org/10.1109/IranianMVIP.2017.8342363

Kumar A, Oishi T, Ono S, Banno A, Ikeuchi K (2013) Global coordinate adjustment of 3D survey models in world geodetic system under unstable GPS condition. 20th ITS World Congress Tokyo 2013

Liang J, Zhang Y, Mei S (2017) Hyperspectral and multispectral image fusion using dual-source localized dictionary pair. In 2017 international symposium on intelligent signal processing and communication systems (ISPACS), pp 261–264. https://doi.org/10.1109/ISPACS.2017.8266485

Liu P, Eom KB (2013) Restoration of multispectral images by total variation with auxiliary image. Opt Lasers Eng 51(7):873–882. Retrieved from http://www.sciencedirect.com/science/article/pii/S0143816613000572. https://doi.org/10.1016/j.optlaseng.2013.02.001

Liu S, Li L, Gao W, Zhang Y, Liu Y, Wang S, Lu J (2018) Diagnosis of nitrogen status in winter oilseed rape (brassica napus l.) using in-situ hyperspectral data and unmanned aerial vehicle (uav) multispectral images. Comput Electron Agric 151:185–195. https://doi.org/10.1016/j.compag.2018.05.026

Lu D, Weng Q (2004) Spectral mixture analysis of the urban landscape in Indianapolis with Landsat ETM+ imagery. Photogramm Eng Remote Sens 70(9):1053–1062

Martins F, Escobar R, Cortes C, Pino A, Salgado M, Pereira E, Cardemil J (2015) Estimating the potential for solar energy utilization in Chile by satellite-derived data and ground station measurements. Sol Energy 121:139–151. https://doi.org/10.1016/j.solener.2015.08.034

Palsson F, Sveinsson JR, Ulfarsson MO (2017) Multispectral and hyperspectral image fusion using a 3-d-convolutional neural network. IEEE Geosci Remote Sens Lett 14(5):639–643. https://doi.org/10.1109/LGRS.2017.2668299

Qian S, Chelikani S, Wang P, Collins LM., Bradbury K, Malof JM (2017) Trading spatial resolution for improved accuracy when using detection algorithms on remote sensing imagery. In 2017 IEEE International Geoscience and Remote Sensing Symposium (IGARSS), pp 3716–3719. https://doi.org/10.1109/IGARSS.2017.8127806

Qin J, Chao K, Kim MS, Lu R, Burks TF (2013) Hyperspectral and multispectral imaging for evaluating food safety and quality. J Food Eng 118:157171. https://doi.org/10.1016/j.jfoodeng.2013.04.001

Restaino R, Vivone G, Dalla Mura M, Chanussot J (2016) Fusion of multispectral and panchromatic images based on morphological operators. IEEE Trans Image Process 25(6):2882–2895

Saritha S, Kumar S (2015) A hierarchical framework for the classification of multispectral imagery. Procedia Comput Sci 46:78–85. https://doi.org/10.1016/j.procs.2015.01.060

Saura JR, Reyes-Menendez A, Palos-Sanchez P (2019) Mapping multispectral digital images using a cloud computing software: applications from UAV images. Heliyon 5(2):e01277. Retrieved from http://www.sciencedirect.com/science/article/pii/S2405844018328184. https://doi.org/10.1016/j.heliyon.2019.e01277

Su J, Liu C, Coombes M, Hu X, Wang C, Xu X, Chen W-H (2018) Wheat yellow rust monitoring by learning from multispectral UAV aerial imagery. Comput Electron Agric 155:157–166. Retrieved from http://www.sciencedirect.com/science/article/pii/S0168169918312584. https://doi.org/10.1016/j.compag.2018.10.017

Sun B, Yuan N, Cao C, Hardeberg JY (2018) Design of four-band multispectral imaging system with one single-sensor. Futur Gener Comput Syst 86:670–679. https://doi.org/10.1016/j.future.2018.04.056

Sylla D, Minghelli-Roman A, Blanc P, Mangin A, Fan-ton Dandon O (2014) Fusion of multispectral images by extension of the pan-sharpening arsis method. IEEE J Sel Top Appl Earth Observ Remote Sens 7:1781–1791. https://doi.org/10.1109/JS-TARS.2013.2271911

Tian L, Du Q, Younan N, Kopriva I (2016) Multispectral image enhancement with extended offset-sparsity decomposition. In: 2016 IEEE international geoscience and remote sensing symposium (IGARSS). IEEE, pp 4383–4386

Wei Q, Bioucas-Dias J, Dobigeon N, Tourneret J-Y (2014) Hyperspectral and multispectral image fusion based on a sparse representation. IEEE Trans Geosci Remote Sens 53:3658–3668. https://doi.org/10.1109/TGRS.2014.2381272

Xiao TG, Weis JA, Gayzik FS, Thomas A, Chiba A, Gurcan MN, Topaloglu U, Samykutty A, McNally LR (2018) Applying dynamic contrast enhanced MSOT imaging to intratumoral pharmacokinetic modeling. Photo-Dermatology 11:28–35. Retrieved from http://www.sciencedirect.com/science/article/pii/S2213597918300120. https://doi.org/10.1016/j.pacs.2018.07.003

Yoshioka H-T, Liu C, Hayashi K (2015) Multispectral fluorescence imaging for odorant discrimination and visualization. Sensors Actuators B Chem 220:1297–1304. https://doi.org/10.1016/j.snb.2015.07.073

Zhou J, Pavek J, Shelton SC, Holden ZJ, Sankaran S (2016) Aerial multispectral imaging for crop hail damage assessment in potato. Comput Electron Agric 127:406–412. https://doi.org/10.1016/j.compag.2016.06.019

Chapter 13
Application of Unmanned Aerial Vehicle (UAV) for Urban Green Space Mapping in Urbanizing Indian Cities

Shruti Lahoti, Ashish Lahoti, and Osamu Saito

Abstract Geospatial data of urban green spaces (UGS) is mostly unavailable for emerging Indian cities, which effects their provisioning, maintenance and monitoring. In absence of spatial data, the strategic vision and comprehensive urban planning to enhance urban environment of cities witnessing transition are compromised. Owing to the direct link between urban environment and wellbeing of urban dweller, it is utmost important to address this gap and support planning process by providing the required data set of desired spatial and temporal scale. Though satellite imagery and remotely sensed data are widely used in environmental studies, for urban greens the spatial resolution of images is inapt. In addition, high cost acts as a barrier towards its wide scale application. Thus, the study reviews current state of literature on application of UAVs for spatial data generation to allow integrated analysis to support urban planning process. A case example of Nagpur city is presented to highlight specific direct applications. The review finds UAVs as cost effective and efficient tool for images with relevant resolution for planners and decision makers. While regulation hinder its wide applicability, the cost component, flexibility, timely monitoring and accessibility weighs UAVs as suitable tool for data collection in urban areas. As use of UAVs in urban areas is still limited, the study recommends more experimentations and trials to arrive at set methodology which could help in mapping of UGS and other qualitative data gathering to support urban planning and urban greening.

Keywords Urban green space · Urbanization · Unmanned aerial vehicle · Thematic mapping · Urban planning

---

S. Lahoti (✉) · O. Saito
United Nations University Institute for the Advanced Study of Sustainability, Shibuya-ku, Tokyo, Japan
e-mail: arshruti.lahoti@gmail.com; saito@unu.edu

A. Lahoti
Independent Researcher, Tokyo, Japan

R. Avtar, T. Watanabe (eds.), *Unmanned Aerial Vehicle: Applications in Agriculture and Environment*, https://doi.org/10.1007/978-3-030-27157-2_13

13.1 Introduction

In developing countries, cities witness special dynamics of urban transition with uneven demographic densities, changing landscape patterns, traffic and congestion and other environmental challenges (Satterthwaite 2008). India is particularly fore-runner with urban population rising up to 300 million by 2050 (UN-Habitat 2016). By 2050, Indian will have 68 cities with population of over a million (Mell 2010), and the urban transition will lead to significant proportions of agricultural land conversion with substantial impacts on their urban environment and loss of green spaces combined with climate change challenges (Govindarajulu 2014). This rapid urbanization is characterized by patterns and process of land use change (Schetke et al. 2016), and poses big challenge towards planning and expansion of urban center while maintaining the urban green space provisions (UGS) in the cities.

UGS contributes towards sustainability of cities by maintaining urban ecosystem and providing range of ecological and social benefits as evidenced in literature. The extensive research on UGS has showcased the diverse and broad range of benefits UGS provides through various ecosystem service provisioning (Kabisch et al. 2015; Tzoulas et al. 2007) as well as cultural services which foster recreation activities, enhance the aesthetics, nurture human nature connection, knowledge sharing and preserve natural landscape features (Priego et al. 2008; Bowler et al. 2010; Lovell and Taylor 2013). This clearly identifies UGS as a crucial part of cities for the livability and the well-being of urban dwellers. However, in developing countries UGS are under stress due to over use and are threatened in transition process, while the vulnerability is further accelerated due to global environmental change (Bhaskar 2012). In India, regardless of the regulation and conservation strategies, the planning, monitoring and management of UGS pose a big challenge (Rao and Puntambekar 2014).

13.2 UGS in Urban Transition

In the emerging urban centers (over one million population) the urban planning efforts are disproportionate as compared to the metropolitan cities, with less priority towards UGS provisions against other infrastructure like housing, water and sanitation, energy supply, which accelerate the challenges. Though the UGS benefits are recognized (Alberti et al. 2003), in general UGS are undervalued and are faced with either destruction or degradation in almost all major cities of India (Rao and Puntambekar 2014). The change in land use with a lack of UGS stewardship is resulting in fragmented and stressed UGS (Dallimer et al. 2015). The biggest challenge faced by authorities and urban planners in this situation is lack of the capacity towards planning and implementing the change (ICLEI -South Asia 2015), due to unavailability of data sets or records of UGS provisions (Town and Country Planning Organisation 2014). This data deficit with rapid transition and unauthorized land conversion makes the monitoring and management of UGS difficult and decline in their proportion in city land use. The dependency of municipal authorities on cadastral maps with lack of details about the dynamic and complex landscape system hinders the decision-making process. Thus, arises the need for more detailed spatial data to guide towards more integrated understanding of UGS (Troy and Wilson 2006; Meyer and Grabaum 2008).

13.3 UGS Thematic Mapping

Human dominated landscape in comparison to natural landscapes deals with constant change and evolves as cities which are dynamic in nature, thus urban planners are faced with several challenges to manage and monitor these dynamic landscapes to maintain the urban environment. In order to plan, maintain and manage these dynamic greens and take informed decision with regard to their provisions, it is important to have efficient technologies and methods to gather topographical information or data. The data would allow planner understand these dynamic landscapes in a comprehensive manner and allow effective translation of strategic visions into strategic action plans (Niluka et al. 2016). The mapped data set is well established in developed countries, while developing countries lack such data set (Sreetheran and Adnan 2007). Like, Greenspace Scotland have spatial data set for 23 different types of green and open spaces at national scale using GIS (Kafafy 2010). Singapore has also implemented their green actions with enhanced green infrastructure city level using spatial data of UGS (Tan et al. 2013). Many European countries are developing comprehensive data set of greens using GIS and other recent technologies for protection and conservation of greens (CABE Space 2010).

In India as well, new methods are applied for mapping the protected greens. Even urbanization and changing land use land cover trends studies are evident (Bhaskar 2012), however, spatial data with finer details is scarce. Owing to the dynamic nature and complex mix of land cover, ground data collection and timely update of data is a tedious task, thus, most of the land use maps are not updated and lack comprehensive record of data. Though it is vital for planning and decision making, at present spatial and digitized data for UGS is not available for most of the Indian cities, thus adversely affecting the planning process. The authors Anguluri and Narayanan emphasized the urgent need of UGS mapping for land allocations during master plan for emerging urban centers (Anguluri and Narayanan 2017). The thematically mapped data with spatial and non-spatial information can facilitate assessment of availability and distribution of typologies of greens prevalent in the city, to equitably evaluate and make decisions between different competing land use demands.

UGS related decision-making regarding land parcels requires information on the spatial distribution of UGS, their functions, vegetation abundance and many more attributes, however, the current cadastral maps include just land use data. The spatial information regarding the landscape functions performed by UGS needs to be added through intensive field work and specially prepared cartographic data, which is not recorded so far. The lack of data set acts as a barrier to effectively manage UGS in urban transition process, thus affecting the overall urban environment in changing land dynamics. Against this background, considering the importance of mapped data in assisting planners and policy makers to design strategies for the optimization of greens, the study reviews potential methods and available technologies used by other disciplines, their application and limitations for mapping UGS to support urban planning process.

13.4 GIS and Remote Sensing

For Spatial data generation, Geographic information system (GIS) is widely accepted as an 'automated systems for the capture, storage, retrieval, analysis, and display of spatial data' (Clarke 1995, p. 13). GIS is very effective in integration of data, representation of data and communication, and to guide urban planning process of UGS. In addition to satellite, data procured from aircraft based remote sensing also provides useful information for monitoring natural resources for a number of years. GIS is widely used to maintain the inventories of urban green space globally. The raster data from satellite is used for the monitoring of vegetation in urban and rural areas (Lang 2008; Liu et al. 2015). The satellite imagery and raster data allow to identify greens consistently across larger geographical areas during the classification process of imagery in form of arbitrary polygons with distinctive land cover, however, the data may not be meaningful to planners in terms of units recognizable on the ground (Yusof 2012). Most of this data classification is based on the reflectance of ground cover, where the unit size and the classes of greens considered are broad and lack finer classes necessary for human managed landscapes.

Also, as compared to natural vegetation (like forest and rangelands), the managed vegetation of urban areas is fragmented and hence more difficult to map with accuracy. Thus, the accurate mapping and extraction of urban vegetation have been critically studied in remote sensing field (Nichol and Lee 2005). For high accuracy the remotely sensed data needs lot of ground survey data in case of human dominated landscape, which is time consuming and hindered by accessibility issue, thus mapping studies in urban areas are very limited. Hence, new spatial data that considers detailed attributes of human dominated landscapes in addition to quantitative green space cover is needed, to allow more sophisticated approaches for analyzing greens for urban planning purpose (Dennis et al. 2018). In addition, the satellite data mismatch in the pixel resolution (30 m) is not suitable to capture the local objectives (Wulder et al. 2004). These limitations are addressed by finer resolution data (10 m resolution) provided by commercially operated satellite sensors like IKONOS and Quickbird. These new sensors are successful in capturing urban vegetation with finer spatial resolution (Johansen et al. 2007) but these images are expensive and sometimes unclear due to cloud cover (Loarie et al. 2007; Anderson and Gaston 2013). Further, the human dominated urban landscapes need frequent monitoring as they are dynamic in nature, thus the cost barrier restricts its wide and local scale application for time series monitoring and assessment. Thus arises the need to explore new methods and techniques to overcome the identified issues and provide data required for land use planning and monitoring purpose.

13.5 Unmanned Aerial Vehicle (UAV)

Use of Unmanned aerial vehicles (UAVs) in civilian areas has recently increased with introduction of cost effective UAV carriers. This new technology has revolutionized the aerial photogrammetric mapping, and its use in various research fields is increasing. As compared to conventional methods, UAV photogrammetry acts as an advanced tool for photographic measurement, where the major components are unmanned aircraft, a transmitter, a communication link, an image sensor like a digital or infrared camera and mission plan (Niluka et al. 2016). UAVs integration with a Light Detection and Ranging (LiDAR) also has wide range of applications (Sofonia et al. 2019). It offers fine scale remote sensing images at a low cost and at required spatial and temporal scale appropriate for urban planning. Thus, this new technology has emerged as an alternative which can be very effective in gathering finer resolution local data to revolutionize landscape ecology and environmental science related research (Anderson and Gaston 2016).

The use of UAVs allows more advanced analysis in comparison to satellite imagery, and also captures ground data efficiently than the conventional ground surveying (Congalton and Green 2009). UAVs has the potential to capture remotely sensed data of high spatial resolution at lower cost with minimum visual intrusion (Noor et al. 2018b; Rango et al. 2009; Raparelli and Bajocco 2019). Further, UAVs can be operated in relatively small clear weather windows with low chances of cloud contamination in the data and offer the flexibility in arranging revisits if needed (Herwitz et al. 2003). The recent literature on UAVs signifies the importance of UAVs in different areas of research like forestry, wildlife studies, crop monitoring and marine investigations (Getzin et al. 2014; Shahbazi et al. 2014; Husson et al. 2014; Bertacchi et al. 2019). The ultra-high resolution (UHR) images received from UAVs allow to assess ground objects in detail, with high accuracy and even identification of mapped vegetation. The satellite data results in land cover classification at coarse scale, while UAVs help to capture the local scale landscape details (Iizuka et al. 2018). The UHR data allows detection of water features, bare soil and other landscape characters and their mapping, while in coarser data such characteristics were generalized and mergers affecting the accuracy of map. Thus, the average accuracy of mapped data though UAVs is higher and the data more reliable and detailed for local landscape information which is important in case of human dominated or managed landscapes.

The review of literature has highlighted some of the key applications of lightweight UAVs in variety of landscape monitoring. The UAVs use in forestry research is for resources assessment, like forest fire monitoring and forest fire recovery (Horcher and Visser 2004; Zajkowski 2003). UAVs are effectively used in ecology, geobotany, natural resource management (Shahbazi et al. 2014; Marcaccio et al. 2016). In rangeland management, the researchers used UAVs to obtain images over small research plots that are harvested to assess the effects of grazing (Quilter and Anderson 2001). The benefit of lightweight UAV is that they can capture data at a low altitudes, thus the individual plant species can be recognized and vegetation dynamics be well captured (Getzin et al. 2012). In the biophysical data, the canopy gap analysis regarding floristic biodiversity can also be recorded (Getzin et al. 2012). All the above-mentioned examples illustrate the applicability of UAVs in natural environmental management and monitoring.

In urban areas, the study by Quanlong et al. used UAV to produce land cover map to help planners and decision makers understand the UGS of cities, however, such studies are still countable and mainly focused on forest and natural vegetation (2015). Another study highlights that low cost commercial UAVs are effective and accurate to capture tree canopy in urban areas and support planning decisions with regard to access to UGS (Niluka et al. 2016). In the field of infrastructure management, a recent study by Saad et al. demonstrates UAVs wide application in mapping ruts and potholes on road surface, thereby helping in monitoring road condition (Saad and Tahar 2019). UAVs are also used as interactive tool to facilitate interaction between community and municipal authority to generate place based solutions, where in UAVs provide open data information for local public and using CAD and other software participatory urban planning among different stakeholders is carried out. UAVs use was also seen in urban design where UAV captured imagery for sites, view of street corridors, images of natural areas and surrounding communities' urban form was collected where the sites were inaccessible (Voigt, as cited in Jenkins 2013). Though UAVs can be very effective for UGS mapping, their potential to capture urban vegetation is not much explored so far and needs attention. The representative examples give some key role played by UAVs, while further exploration on methods is required to realize its full fledge applications in UGS mapping and monitoring (Goodman 2014).

13.6 Case Example of UAVs Use to Support Urban Planning by Mapping UGS

Nagpur city is the 13th largest city in India, with a spread of 217.65 km^2 and population of 2.4 million (as of 2011 census). The city holds the title of greenest city, however, the city is witnessing socio-economic changes, land use changes and expansion of city boundaries with demographic rise in last decade. In this transition process, the urban environment of the city is degrading with increased air and water pollution, decline in green spaces, wetlands, ground water depletion and reduction in water infiltration, increased flash floods, and increased expansion and urban heat island effect (Dhyani et al. 2018). The greenest city is sprawling with substantial loss and degradation of natural greens (Surawar and Kotharkar 2012, 2017; Lanjewar and

Kelkar 2007; Verma et al. 2011), and in absence of baseline data the authorities are facing challenges in realizing a comprehensive vision towards maintenance and monitoring of urban environment of the city. This highlights the urgent need to overcome the data deficit by mapping and creating spatial data to guide planning process.

Also, the thematic map of greens is needed owing to the recent policy innovations and reforms of the Government of India (GOI) under Atal Mission for Rejuvenation and Urban Transformation (AMRUT). As per AMRUT, all emerging urban centers are required to prepare Service Levels Improvement Plans (SLIPs) to assess the service level gaps in infrastructure provisions, based on which the new master plans need to be prepared (Ministy of Urban Development 2015a). Thus, under SLIPs the municipal authorities are asked for primary assessment "the existing situation and service levels gaps for organized Green Space and Parks based on standards prescribed in Urban and Regional Development Plan Formulation Implementation (URDPFI) Guidelines" (Ministry of Urban Development 2015b). However, for very few cities the assessment is available and for Nagpur the benchmarks are far below the bottom line standards (Lahoti et al. 2019a). Even though this is an important data for planning and managing future UGS, the municipal authorities struggle to have the consolidated spatial data of public UGS and the insufficiency of this data hinders detailed assessment and related planning process (Lahoti et al. 2019b).

In such scenario, UAVs with highly adaptive methods prove as a potential tool for planners with ability to capture aerial data by which they can prepare the much needed data set. Being inexpensive, the tool can provide access to data and information which was earlier unavailable due to cost factor for many planning tasks. Moreover, the adaptability of the technology as well as the cost components favors utilization of UAVs in urban planning for data collection task. The above reviewed literature highlights UAVs as data-gathering tools with several benefits like cost savings, flexibility, less cloud error, time series monitoring and time savings. Among these, the biggest incentive for adopting the technology is the unique and accurate data (Jenkins 2013), which can be very effective in evaluating the current benchmarks fulfilled by Municipal authorities (as per SLIP) reliably and affordably, while revisiting the parameters in a timely fashion in transition process. The UAVs can assist the municipal authorities to overcome the data deficit issue.

Also, UAVs can assist in urban vegetation mapping, where extraction of vegetation data and identification of vegetation can be done though acquired imagery (Noor et al. 2018a). Also, other attributes like surfaces, street furniture, canopy cover, park entrance, water feature and many others attributes depending upon the mapping objective can be partially derived from UAV data. These objects can be digitized and can be later updated or verified through site inspection. Other data can be retrieved from UAV images by advanced processing, like species identification, which needs disciplinary and local knowledge. This geospatial data (with added attributes) about the land parcel can be useful in decision making with added inputs through local information and field inspections (Noor et al. 2018a). Other invisible data sets can be further added to this spatial data like physical boundaries, population tracts to carry out benchmark analysis of per capita available greens. The retrieved data is useful for planners and policy makers to understand the service area gap in comparison to the bottom line standards. Further, qualitative data like location of parks along with their entrance gates, surrounding road network, facilities, and the neighborhood structure can help in evaluation of UGS provision. UAVs also allow qualitatively data collection by effective and efficient ground surveying (Watts et al. 2010), thus a potential tool to assimilate data (quantitative and qualitative) to allow comprehensive understanding of current status of greens, their supply side gap, the derived benefits which can help planners and policy makers for informed decision making and their prioritization and implementation as part of broader infrastructure.

13.7 Conclusion

In the urban transition scenario prevalent in India, where urban greens are largely affected due to lack of comprehensive vision towards urban greening, there is an immediate need to support urban planning process by providing the required data set of desired spatial and temporal scale. Against this background, the study highlights the use of UAVs as a cost effective and efficient tool for more integrated data recording of UGS in emerging urban centers of India. The unprecedented potential of small UAVs as a low altitude imaging platform for geospatial information gathering is widely recognized by many researchers. UAVs are apt to complement the existing technologies by overcoming the identified shortcomings with regard to UGS mapping. Based on the literature review, the study recommends more experimental research and exploration of spatial data generation methods using GIS techniques and UAVs to assess and monitor UGS both quantitatively and qualitatively. Though UAVs use in geospatial information acquisition is omnipresent, their potential to map UGS is still widely applied. The case example illustrates various ways in which UAV procured High resolution image (HRI) and ground survey data can help in generating thematic maps of UGS. Use of UAVs in mapping can help to overcome the data deficit, wherein the retrieved and translated data can be used to guide urban planning related decisions.

Furthermore after baseline data generation with detailed and local level spatial data, UAVs along with open source photogrammetric software can allow community driven UAV mapping activities in future. Despite ethical implications, complex legislative process and stringent regulation with regard to permission (Noor et al. 2018a; Carey 2014; Rajagopalan and Krishna 2018), the manifold advantages in terms of the turnaround time for data set generation, low cost component, high resolution spatial data at local scale make them most suitable tool for urban areas, where greens are subjected to change frequently due to continuous anthropogenic activities. Also, more attention is required towards evaluation of data processing techniques in addition to hardware development (Shahbazi et al. 2014), thus, more experimental research is suggested in order to effectively apply UAVs as mainstream tools for UGS mapping, monitoring and management.

Conflict of Interest The authors declare that they have no conflict of interest.

References

Alberti M, Marzluff JM, Shulenberger E, Bradley G, Ryan C, Zumbrunnen C (2003) Integrating humans into ecology: opportunities and challenges for studying urban ecosystems. Bioscience 53(12):1169–1179

Anderson K, Gaston JK (2013) Lightweight unmanned aerial vehicles will revolutionize spatial ecology. Front Ecol Environ 113:138–146. https://doi.org/10.1890/120150

Anguluri R, Narayanan P (2017) Role of green space in urban planning: outlook towards smart cities. Urban For Urban Green 25:58–65

Bertacchi A, Giannini V, Di C, Nicola F (2019) Using unmanned aerial vehicles for vegetation mapping and identification of botanical species in wetlands. Landsc Ecol Eng. https://doi.org/10.1007/s11355-018-00368-1

Bhaskar P (2012) Urbanization and changing green spaces in Indian cities (case study – city of Pune). Int J Geol Earth Environ Sci 2-2:148–156

Bowler D, Buyung-Ali L, Knight T, and Pullin AS (2010) The importance of nature for health: is there a specific benefit of contact with green space? Environmental Evidence. Retrieved from: https://www.cabdirect.org/cabdirect/abstract/20103229510

CABE Space (2010) Urban green nation: building the evidence base. CABE Space, London

Carey B (2014) Eurpean commission vows tough regulation of UAVs. Retrived from: https://www.ainonline.com/aviation-news/2014-04-10/european-commission-vows-tough-regulation-uavs

Clarke K (1995) Analytical and computer cartography, 2nd edn. Prentice Hall, Upper Saddle River

Congalton RG, Green K (2009) Assessing the accuracy of remotely sensed data; principles and practices, 2nd edn. CRC/Tylor and Francis, New York

Dallimer M, Davies ZG, Diaz-porras DF, Irvine KN, Maltby L, Warren PH et al (2015) Historical influences on the current provision of multiple ecosystem services. Glob Environ Chang 31:307–317

Dennis M, Barlow D, Cavan G, Cook P, Gilchrist A, Handley J et al (2018) Mapping urban Green infrastructure: a novel landscape-based approach to incorporating land use and land cover in the mapping of human-dominated systems. Land 7(1):17. https://doi.org/10.3390/land7010017

Dhyani S, Lahoti S, Khare S, Pujari P, Verma P (2018) Ecosystem based Disaster Risk Reduction approaches (EbDRR) as a prerequisite for inclusive urban transformation of Nagpur City, India Int J Disaster Risk Reduct. https://doi.org/10.1016/j.ijdrr.2018.01.018

Getzin S, Wiegand K, Schoning I (2012) Assessing biodiversity in forest using very high-resolution images and unmanned aerial vehicles. Methods Ecol Evol 3:397–404

Getzin S, Nuske RS, Wiegand K (2014) Using unmanned aerial vehicle (UAV) to quantify spatila gap patterns in forest. Remote Sens 6:12815–12836

Goodman P (2014) Drone mapping for urban planning in Mexico City. [ONLINE] Available at: https://www.mapbox.com/blog/flying-mexico-city/. [Last Accessed 10/4/2019]

Govindarajulu D (2014) Urban green space planning for climate adaptation in Indian cities. Urban Clim 10(P1):35–41. https://doi.org/10.1016/j.uclim.2014.09.006

Herwitz SR, Johnson L, Dunagon S, Higgins R, Sullivan D, Zheng J, Witt G (2003) Cofee field ripeness detection using high resolution imaging systems on a solar powered UAV. In: Proceedings of 30th symposium on Remote Sensing of Environment, 12.3–3. Honolulu

Horcher A, Visser RJ 2004) Unmanned aerial vehicles: application for natural resource management and monitoring. In: Proceedings of Council Forest Engineer. Annual meeting–machines and people, the interface, Hot Springs, AR, 5 pp

Husson E, Hagner O, Ecke F (2014) Unmanned aircraft systems help to map aquatic vegetation. Appl Veg Sci 17(3):567–577. https://doi.org/10.1111/avsc.12072

ICLEI-South Asia (2015) Green growth good practices. Delhi, India: ICLEI-Local Governments for Sustainability, South Asia (vol 3)
Iizuka K, Masayuki I, Satomi S, Takashi M, Mark D, Watanabe K (2018) Advantages of unmanned aerial vehicle (UAV) photogrammetry for landscape analysis compared with satellite data: a case study of postmining sites in Indonesia. Cogent Geosci 4:1498180
Jenkins NAL (2013) An application of aerial drones in zoning and urban land use planning in Canada a preliminary review of current policies, restrictions and planning direction for aerial drones in Canadian cities. University of Guelph
Johansen K, Coops NC, Gergel SE, Stange Y (2007) Application of high spatial resolution satellite imagery for riparian and forest ecosystem classification. Remote Sens Environ 110:29–44
Kabisch N, Qureshi S, Haase D (2015) Human-environment interactions in urban green spaces – a systematic review of contemporary issues and prospects for future research. Environ Impact Assess Rev 50:25–34
Kafafy N (2010) The dynamics of urban green space in an arid aity. The case of Cairo-Egypt. PhD dissertation. Cardiff University, UMI Number: U584501
Lahoti S, Kefi M, Lahoti A, Saito O (2019a) Mapping methodology of public urban green spaces using GIS: an example of Nagpur City, India. Sustainability 11:1–23
Lahoti S, Lahoti A, Saito O (2019b) Urban Forestry & Urban Greening Benchmark assessment of recreational public urban green space provisions: a case of typical urbanizing Indian City, Nagpur. Urban For Urban Green 44:126424. https://doi.org/10.1016/j.ufug.2019.126424
Lang S (2008) Object-based image analysis for remote sensing applications: modelling reality – dealing with complexity. In: Blaschke T, Lang S, Hay GJ (eds) Object based image analysis. Springer, New York/Heidelberg/Berlin, pp 1–25
Lanjewar KY, Kelkar PS (2007) Studies on rejuvenation of Futala Lake, Nagpur. In: The 12th world lake conference, pp 1660–1668. Retrieved from http://www.moef.nic.in/sites/default/files/nlcp/P-WorldCase Studies/P-44.pdf
Liu Y, Meng Q, Zhang J, Zhang L, Jancso T, Vatseva R (2015) An effective Building Neighborhood Green Index model for measuring urban green space. Int J Digit Earth. Taylor and Francis. https://doi.org/10.1080/17538947.2015.1037870
Loarie SR, Joppa LN, Pimm SL (2007) Satellite miss environmental priorities. Trends Ecol Evol 22:630–632
Lovell ST, Taylor JR (2013) Supplying urban ecosystem services through multifunctional green infrastructure in the United States. Landsc Ecol 28(8):1447–1463. https://doi.org/10.1007/s10980-013-9912-y
Marcaccio JV, Markle CE, Chow-Fraser P (2016) Use of fixed-wing and multi-rotor unmanned aerial vehicles to map dynamic changes in a freshwater marsh. J Unmanned Veh Syst 43:193–202. https://doi.org/10.1139/juvs-2015-0016
Mell IC (2010) Green infrastructure: concepts, perceptions and its use in spatial planning. Landscape, (June), 290. Retrieved from http://www.academia.edu/9517808/Green_Infrastructure_concepts_perceptions_and_its_use_in_planning
Meyer BC, Grabaum R (2008) MULBO—model framework for multi critieria landscape assessment and optimisation. A support system for spatial land use decisions. Landsc Res 33:155–179
Ministry of Urban Development (2015a) Smart cities: Mission Statement & Guidelines, government of India. https://doi.org/10.1016/S0264-2751(98)00050-X
Ministry of Urban Development, (2015b) Atal Mission for Rejuvenation and Urban Transformation (AMRUT). Retrieved from http://amrut.gov.in/writereaddata/AMRUTGuidelines.pdf
Mohd Noor N, Abdullah A, Hashim M (2018a) Evaluating the societal impact of using drones to support urban upgrading projects. In: IOP conference series: earth and environmental science (vol 169). https://doi.org/10.3390/ijgi7030091
Mohd Noor N, Abdullah A, Hashim M (2018b) Remote sensing UAV/drones and its applications for urban areas: a review. IOP Conf Ser Earth Environ Sci 169. https://doi.org/10.1088/1755-1315/169/1/012003

Nichol J, Lee CM (2005) Urban vegetation monitoring in Hong Kong using high resolution multispectral images. Int J Remote Sens 26(5):903–918

Niluka M, Uideshini M, Premasiri HMN, Madawalagama S, Samarakoon L et al (2016) Developing methodology to map tree canopy in urban areas from low cost commercials UAVs. In: 37th Asian conference of remote sensing, Colombo(37). Retrieved from https://www.researchgate.net/publication/309349770_DEVELOPING_METHODOLOGY_TO_MAP_TREE_CANOPY_IN_URBAN_AREAS_FROM_LOW_COST_COMMERCIAL_UAVS

Priego C, Breuste JH, Rojas J (2008) Perception and value of nature in urban landscapes: a comparative analysis of cities in Germany, Chile and Spain. Landsc Online 7:1), 1–1),22. https://doi.org/10.3097/LO.200807

Quilter MC, Anderson VJ (2001) A proposed method for determining shrub utilization using (LA/LS) imagery. J Range Manag 54:378–381

Rajagopalan R, Krishna R (2018. Drones: guidelines, regulations, and policy gaps in India. OrfonlineOrg. Retrieved from http://www.orfonline.org/research/drones-guidelines-regulations-and-policy-gaps-in-india/

Rango A, Laliberte A, Herrick J, Winters C, Havstad K, Steele C, Browning D (2009) Unmanned aerial vehicle-based remote sensing for rangeland assessment, monitoring, and management. J Appl Remote Sens 3(1):33542. https://doi.org/10.1117/1.3216822

Rao P, Puntambekar K (2014) Evaluating the urban green space benefits and functions at macro, meso and micro level: case of Bhopal City. Int J Eng Res Technol 3:359–369

Raparelli E and Bajocco S (2019) A bibliometric analysis on the use of unmanned aerial vehicles in agricultural and forestry studies. Int J Remote Sens, 0(0), 1–14. https://doi.org/10.1080/01431161.2019.1569793

Saad AM, Tahar KN (2019) Identification of rut and pothole by using multirotor unmanned aerial vehicle (UAV). Measurement 137:647–654. https://doi.org/10.1016/j.measurement.2019.01.093

Satterthwaite D (2008) Cities' contribution to global warming: notes on the allocation of greenhouse gas emissions. Environ Urban 20:539–549. https://doi.org/10.1177/0956247808096127

Schetke S, Qureshi S, Lautenbach S, Kabisch N (2016) What determines the use of urban green spaces in highly urbanized areas? - examples from two fast growing Asian cities. Urban For Urban Green 16:150–159. https://doi.org/10.1016/j.ufug.2016.02.009

Shahbazi M, Théau J, Ménard P (2014) Recent applications of unmanned aerial imagery in natural resource management. GIScience Remote Sens 51(4):339–365. https://doi.org/10.1080/15481603.2014.926650

Sofonia JJ, Phinn S, Roelfsema C, Kendoul F, Rist Y (2019) Modelling the effects of fundamental UAV flight parameters on LiDAR point clouds to facilitate objectives-based planning. ISPRS J Photogramm Remote Sens 149:105–118. https://doi.org/10.1016/j.isprsjprs.2019.01.020

Sreetheran M, Adnan, M (2007) Green network development of Kuala Lumpur City: From the perspective of Kuala Lumpur Structure Plan. FORUM Int J Postgraduate Studies Architecture Planning Landscape 7(1):38–41

Surawar M, Kotharkar R (2012) Environmental impact assessment in urban fringe to achieve urban sustenance. In: International conference on advances in architecture and civil engineering (AARCV 2012), vol 2, pp 767–774

Surawar M, Kotharkar R (2017) Assessment of urban Heat Island through remote sensing in Nagpur urban area using Landsat 7 ETM+. Int J Civil Environ Struct Construct Architect Eng 11(7):851–857

Tan PY, Wang J, Sia A (2013) Perspectives on five decades of the urban greening of Singapore. Cities 32:24–32

Town and Country Planning Organisation (2014) Urban greening guidelines. Retrieved from http://www.indiaenvironmentportal.org.in/files/file/urbangreenguidelines2014.pdf

Troy A, Wilson MA (2006) Mapping ecosystem services: practical challenges and opportunities in linking GIS and value transfer. Ecol Econ 60:435–449. https://doi.org/10.1016/j.ecolecon.2006.04.007

Tzoulas K, Korpela K, Venn S, Yli-Pelkonen V, Kaźmierczak A, Niemela J, James P (2007) Promoting ecosystem and human health in urban areas using green infrastructure: a literature review. Landsc Urban Plan 81(3):167–178. https://doi.org/10.1016/j.landurbplan.2007.02.001

UN-Habitat (2016) World cities report 2016. Urbanization and development: emerging futures

Verma SR, Chaudhari PR, Singh RK, Wate SR (2011) Studies on the ecology and trophic status of an urban Lake at Nagpur City, India. Rasayan J Chem 4(3):652–659

Watts AC, Perry JH, Smith SE, Burgess MA, Wilkinson BE, Szantoi Z et al (2010) Small unmanned aircraft systems for low-altitude aerial surveys. J Wildl Manag 74(7):1614–1619. https://doi.org/10.2193/2009-425

Wulder MA, Coops NC (2014) Satellites: make earth observations open access. Nature 513:30–31. https://doi.org/10.1038/513030a

Yusof MJM (2012) The true colours of urban green spaces: identifying and assessing the qualities of Green spaces in Kuala Lumpur, Malaysia. Doctoral thesis, Institute of Geography, University of Edinburgh

Zajkowski TJ (2003) "Unmanned aerial vehicles: remote sensing technology for the USDA Forest Service," project report RSAC-1507-RPT1, 15 pp., rem. Sens. Applications Center, Salt Lake City, UT

Chapter 14
Responsibility and Accountability in the Governance of Civilian UAV for Crop Insurance Applications in India

Anjan Chamuah and Rajbeer Singh

Abstract In an attempt to govern civilian unmanned aerial vehicle (UAV) for crop insurance applications, the paper is considering the risk and challenges associated with the emerging technology to be deployed in Indian agriculture, which is again highly diversified with varied physical features and socio-cultural practices of the agriculture community. These issues of governance concerning accountability and responsibility of actors and institutions are leading research problems of the study and would be addressed as research questions – **How are the challenges of governance of civilian UAV innovations in crop insurance application can be addressed? How can the responsibility and accountability of the actors ensure effective governance?** As such, the paper draws empirical results from in-depth interviews carried out as a part of the primary survey based on snowball sampling technique. Accordingly, the paper advances in adhering to the responsible deployment of the technology and ushering accountability in governance to enhance civil UAV innovations in the crop insurance application. Besides, institutional arrangements which help proper regulations of the technology, upholding values such as transparency, trust, privacy, effectiveness and efficiency can enhance an effective governance structure for civil UAV innovations in crop insurance applications.

Keywords Civilian UAV · Governance · Accountability · Responsibility · Innovation

A. Chamuah (✉)
Centre for Studies in Science Policy, School of Social Sciences, Jawaharlal Nehru University, New Delhi, India

R. Singh
Centre for Studies in Science Policy, School of Social Sciences, Jawaharlal Nehru University, New Delhi, India

Technical University Eindhoven, Eindhoven, Netherlands

R. Avtar, T. Watanabe (eds.), *Unmanned Aerial Vehicle: Applications in Agriculture and Environment*, https://doi.org/10.1007/978-3-030-27157-2_14

14.1 Introduction

Civilian unmanned aerial vehicle (UAV) is an innovation in Indian agriculture application. The advent of UAV has revolutionised the agriculture sector. Civil UAV mounted with many features and applications has made remote sensing an easier task, which was earlier a herculean job primarily dependent on satellite imagery. However, the arrival of technology also poses many challenges and risk in the agriculture sector. The challenges are related to governance of the technology regarding issues such as accountability, efficiency, trust, privacy, transparency, security and responsibility of the new and emerging technology. New and emerging technology always pose a risk and unknown consequences. Emerging technology is radically novel and relatively fast-growing which has the potential to exert an impact on the economy and society (Rotolo et al. 2015) and is also trying to provide solutions to problems (Nordmann 2014) of human life and their surrounding ecosystem.

Further, with innovations, new technologies are penetrating to the core of our lives (Giddens 1999), from decision making to human resource is affected by new technologies (Mobegi et al. 2012). Innovation is a process; consisting of a linear chronological flow of predefined stages: idea generation, idea selection (screening), development, and launch to the market (Salerno et al. 2015). Innovation is not a linear model with a simple path; instead, it is more complex and dynamic. There are many actors in an innovation process with the zigzag path and tracing accountability is also a humongous task (Owen et al. 2013). However, the challenge is for the governance of the innovation, to what extent it can allow and to whom. In a democratic country like India, with diverse socio-cultural aspects these questions of governance at times pose challenges for the government. Proper evaluation of the consequences of the technology like civilian UAV to the society, economy and environment can only conclude whether the technology is sustainable or not.

The responsible deployment of the technology can address the issues related to the governance of the technology by considering the varied socio-cultural aspect of Indian civilisation. The disposition of the technology should not breach the culture-specific values of society. In a diverse country like India, where the majority of the population is dependent on agriculture ensuring their cultural norms and practices while deploying a new and emerging technology like civil UAV determines the fate of the innovations. However, for effective governance of the technology; responsibility and accountability of all the actors and institutions are also very crucial. These issues of governance concerning accountability and responsibility of actors and institutions are leading research problems of the study and would be addressed as research questions – **How are the challenges of governance of civilian UAV innovations in crop insurance application can be addressed? How can the responsibility and accountability of the actors of innovation ensure effective governance?** These are some of the pertinent questions which would be dealt within the paper.

The rest of the paper is organised as follows: Section 14.2 describes civilian UAV. Section 14.3 addresses the governance issues associated with the regulation of UAV technology in India. Section 14.4 provides an overview of the issue of responsibility and accountability embedded on UAV or drone and how it is deployed in the crop insurance scheme. Section 14.5 discusses the methodology adopted to carry out the research study, whereas Sect. 14.6 illustrates how values are embedded in the deployment of civil UAV and in Sect. 14.7 the paper is concluded.

14.2 Defining a Civilian UAV

UAV or drone is an innovation which has a wide-ranging application. Drones which are used for the civilian purpose are called civilian UAV or drone. Whereas, drones used by the military for defence and war are called military drones. Civilian UAV has a wide-ranging application in agriculture, insurance, energy and utilities, infrastructure, mining, media and entertainment. However, the focus of the paper is on civilian UAV usage on crop insurance applications in India.

A pilotless aircraft is known as a drone or a UAV. The International Civil Aviation Organization (ICAO) which is the nodal agency for controlling drone regulations worldwide defines UAV as:

A UAV is a pilotless aircraft, in the sense of Article 8 of the Convention on International Civil Aviation, which is flown without a pilot-in-command on-board and is either remotely and fully controlled from another place (ground, another aircraft, space) or programmed and fully autonomous (ICAO 2005).

The UAV consisted of different parts; viz., a single board computer, a remote control (RC) aircraft, an inertial measurement unit (IMU), a multi-spectral camera, a wide area augmentation system (WAAS), a global positioning system (GPS), a flight controller, a pulse width modulation (PWM) switch, a video transmitter and a wireless router (Xiang and Tian 2011). The flight control unit (FCU) is the central part of any UAV which is capable of processing the information given by a remote pilot into the navigational task (Siebert and Teizer 2013). The FCU is directly connected to a GPS receiver and a magnetic compass. Control functions for a UAV are either onboard or remotely controlled from the ground. UAV uses aerodynamic forces to provide vehicle lift and can fly autonomously or with the aid of a remote pilot (Peterson 2006). A fixed winged aircraft requires a runway for take-off and land (Valavanis 2007). A UAV carries various types of payloads which enhances its capability of solving specific tasks within and beyond the earth's atmosphere for a specified time of the targeted mission (Peter V Blyenburgh 1999). Different varieties of UAVs are available nowadays; small sized like an insect or a big charter flight. UAVs are large enough to accommodate cameras, sensors and other equipment required to send information to the ground (Finn and Wright 2012). Compiling all its payloads and units, UAV is seen as a system and called as Unmanned Aircraft System (UAS). UAS consists of an Unmanned Aircraft (UA), a Remote Pilot System (RPS), Command and Control (C2) link, the

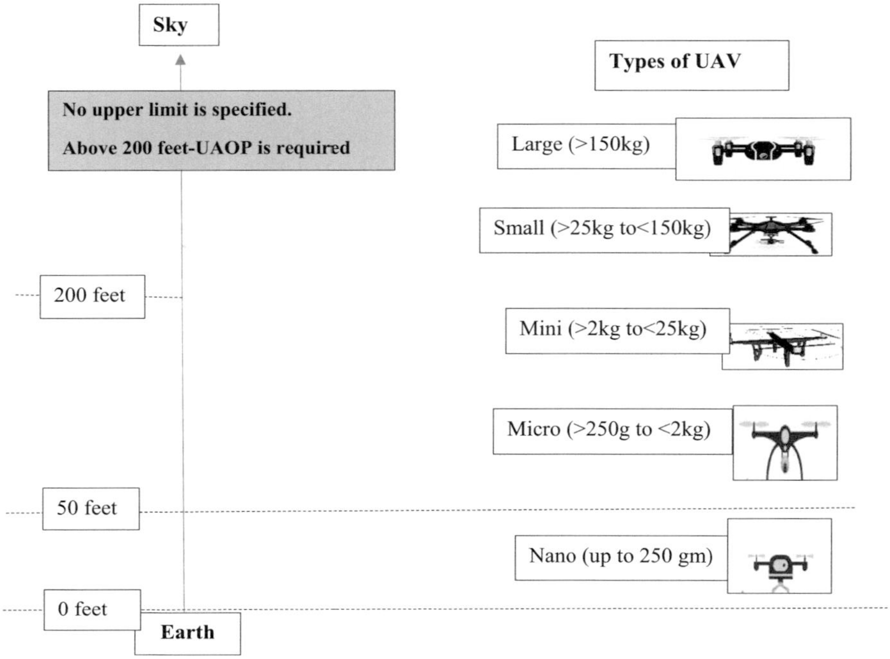

Fig. 14.1 Civilian UAV and its types. (Source: Own compilation from DGCA 2018)

maintenance system and operating personnel (DGCA 2016). Different varieties of civilian UAVs are deployed in agriculture insurance. DGCA has categorised civil UAV into four categories by their weight which is shown in the Fig. 14.1.

It is evident from the figure above that there are five categories of drones in India; Nano, micro, mini, small and large. A Nano drone is allowed to fly to 50 feet above the ground. The other four categories can fly above 50 feet without any restrictions. However, to fly above 200 feet the micro, mini, small, and large drones require to have an Unmanned Aircraft Operator Permit (UAOP) granted by DGCA. All five categories of the drone while flying on the sky should always maintain a Visual Line of Sight (VLOS). The regulations in India thus gives freedom and autonomy to fly above 200 ft, provided it does not enter restricted zones. However, at times, the drones might cross the VLOS and connection with the operator might fail and can lead to collisions and accidents. Thus, it might cause harm leading to unhealthiness, inefficiency, and lack of trust in the drones and the operators.

The next subsection illustrates the governance structure of civil UAV in India, the main actors and institutions governing it are also elaborated.

14.3 Civil UAV Governance in India

Effective governance of emerging technology can lead to responsible innovation. Governance refers to a set of actors and institutions that are drawn from but also beyond government (Stoker 1998). Governance is thus a set of rules and regulations for the authoritative organisation of collective life which has the potential to empower citizens through participation and consultation process (Devaney 2016). Governance enhances the relations of responsibility (Pellizzoni 2004). Governance of emerging technology should be viewed from the historical context based on some flexibility in new developments (Von Schomberg 2013); coordinated between the regulator and the regulated (Lee and Petts 2013). Several aspects like cooperation, collaboration, accountability and transparency are essential for UAV technology governance (Mohamed et al. 2018). There are two dimensions of governance structure; first, is political concerning qualified distributions of public goods for the common good and the second is technically related to capacity building (Ahrens 2002).

Responsible governance of new and emerging technology is possible only after soliciting societal needs, caring institutions, actors and policies that govern (van Oudheusden 2014). Only good governance can anticipate whatever good or bad can happen with new and emerging technology (Nordmann 2014) which has become a testing ground for new approaches to governance (Guston 2012). In the discussion of governance of emerging technology, governance and accountability are interlinked. Here, governance is not always about the day-to-day running of operations; preferably it is about vision, oversight, process, independence and accountability (Jepson 2005).

In India, many actors and stakeholders are involved in UAV governance. Actors here refers to both firm and non-firm actors (Malerba 2002) who perform a specific function within a system (Lelea, Universität Kassel, and Wissenschaftliche Betriebseinheit Tropenzentrum 2014). Stakeholders are persons, groups or organisation which must be taken into consideration by leaders, managers and front-line staff (Bryson 2004) because they take a common interest or stake (Lelea, Universität Kassel, and Wissenschaftliche Betriebseinheit Tropenzentrum 2014) during an innovation process. Director General of Civil Aviation (DGCA) is the nodal agency in monitoring, granting permission and assessment of emerging technology like UAS. Apart from DGCA, other actors of deployment of civilian UAV found by this study are farmers, insurance companies, scientific organisations like Mahalanobis National Crop Forecast Centre (MNCFC), North East Centre for Technology Assessment and Research (NECTAR), Skymet and local administration. Apart from these, some private drone making companies are also engaged in collaboration with government bodies.

Flying a UAV is not an easy task. Various formal institutions (North 1991) regulate its operation. To operate a UA in India, one needs to get a Unique Identification Number (UIN) provided by the DGCA. All Civil UA operation above 200 ft above ground level (AGL) in uncontrolled airspace requires UAOP from DGCA. The validity of the UAOP is for 2 years and is not transferable (DGCA

2016). The UAS issued with UIN should not be sold to any person without the permission of DGCA.

The next subsection describes the dimensions of responsibility and accountability which are crucial for effective governance of new and emerging technology.

14.4 Responsibility and Accountability in Civilian UAV Deployment

There are many challenges in deployment and governance of civil UAV in Indian crop insurance applications. Apart from varied socio-cultural aspects; India has diverse physical conditions and cropping seasons which require an acumen in the deployment of new and emerging technology like civil UAV. Henceforth, the governance of such technology also requires upholding values like accountability and responsibility at the level of the innovation process. Accountability ensures answerability for undesirable action or duties (Mishra and Singh 2018). Similarly, responsibility means accountability and liability for the consequences of his/her actions (Pennock 1952; Swierstra and Jelsma 2006). Accordingly, governance accountability is the answerability of the officials to the other stakeholders on the dispersal of their powers, duties and decisions; accept the responsibility for failure, and incompetencies (Devaney 2016). Thus, when the civil UAV is deployed, it should be socially acceptable, economically affordable and environmentally viable to adhere to the values of accountability and responsibility.

The study is trying to see the deployment of civilian UAV technology in crop insurance in India. Indian Government has launched a programme called "KISAN" in October 2015 (Crop Insurance using Space Technology and Geoinformatics). In this scheme, civilian UAV technology is used for crop insurance. KISAN's use of UAVs is in the experimental stage. Agricultural analysts can use the data gathered through it, and further help the farmers and insurance companies to improve crop insurance in many ways (Garg 2016). Collected data through UAV and gathered images through satellite would be further matched with the earlier adopted crop cutting experiments to arrive at an accurate record (Parsai 2015). The basic principle underlying crop insurance is that many in an area share the loss incurred by a few. The incidence of risk is distributed to those who can bear the risk (Aurora and Rachuri 2016). The images and videos provided by civilian UAV help in payment of claims to farmers during crop loss and damages. Civilian UAV has bridged the gap between farmers and insurance companies. The technology has become a capable platform for communication for agricultural losses and claims.

The advent of UAV technology through the 'KISAN' scheme has revolutionised agriculture insurance in India. The technology with its sensors and high-resolution cameras is capable of providing information unhindered in adverse weather condition. The modern, cost-effective sensors mounted on a UAV provide precise information with lesser complexity. Apart from it, the spectral reflectance and thermal emittance properties of soil and crops can also be evaluated through a UAV (Sinha et al. 2016). UAV helps in acquiring, processing, analysing and managing critical elements of the agricultural production system. Crop insurance thus helps in socialising of losses of farmers, further ensuring accountability, responsibility, transparency and trust with the insurer embedded by UAV innovations.

The next subsection on methodology describes how the research is done; methods, tools and technique deployed in primary and secondary data collection.

14.5 Methodology

The study on civil UAV governance in the crop insurance application is an exploratory one seeing the newness of the technology deployed to Indian agriculture. An extensive literature survey is done to gather theoretical understandings on UAV, issues of governance, accountability and responsibility based on a literature survey questionnaire. The relevant literature is reviewed concerning the research objectives of the study. In-depth interview was carried out at MNCFC, DGCA, NECTAR, and Skymet. All the interviews were conducted personally through face-face interview method. These organisations are based in Delhi. The interviews for the study was conducted from March 2017 to November 2018.

The interview schedule used for the in-depth interviews consisted of a set of open-ended questions. None of the questions in the interview schedule was leading but was probing. The language of the interview schedule was English. Snowball sampling technique was used to identify the interviewee and to get a response to the interview schedule. Snowball sampling is a non-random sampling technique used when characteristics to be possessed by samples are rare and difficult to find (Dudovskiy n.d.). The civilian UAV is a new and emerging technology in Indian agriculture; users and developers are not easy to find, so the snowball technique was adopted. The collected data were transcripted and then analysed according to the stated objectives and research questions of the study.

The next subsection describes values enshrined to civil UAV deployment furthering into the sustainability of the innovation.

14.6 Embedding Values to Civil UAV Deployment

Responsible civil UAV deployment requires adhering to values which are specific to the society where the technology is implemented. Values can be universal or culture-specific (Axelsson et al. 2013; Setiawan and Singh 2015). Universal values, which are common and are given similar consideration across different societies. Cultural-specific, which represents the morals, aspiration, and ethos of a particular society while dealing with innovation. Values are thus indicators or philosophical representation of specific desired functionalities, effects and benefits (Zahinos et al. 2013).

Table 14.1 Values contributing to sustainability dimensions

Values	Variables	Sustainability dimensions
Transparency	Openness in communication	Promotes social sustainability
	Enhances public scrutiny and visibility	
Privacy	Freedom from intrusion, control of information, and surveillance	Promotes social sustainability
	Adhering to ethics and morality of the target population	
Affordability	Ability to bear the cost of any product or services	Promotes economic sustainability
Autonomy	Freedom of control of technology to the user	Promotes social and economic sustainability
	Freedom of self-governing	
Effectiveness	The degree of completion of a task	Promotes economic sustainability
Trust	Expectations from the technology	Promotes social sustainability
	Human reliance on technology	
Safety	The absence of risk and hazards	Promotes social and environmental sustainability
Security	The state of being free from threat and danger	Promotes social and environmental sustainability

The values thus help in philosophising the technological and innovational approach. The role of values is mainly focussed on the means to achieve aims (J. Gonzalez 2015). However, values are not embedded within a technology; somewhat they are implicated through engagement (Hoven et al. 2015). The aim of values is securing sustainability (social, economic and environmental). So, in any governance, our goal is to ensure sustainability and viability. It should not have any adverse impact on the society and environment. Similarly, any law which is coming should not have any impact on innovations.

The study is focused on the culture-specific values of Indian society where the technology is deployed. Values like transparency, privacy, affordability, autonomy, effectiveness, trust, safety and security assume high significance for a new and emerging technology like civil UAV when deployed in Indian agriculture. How these values are related to civil UAV in Indian crop insurance is shown above in Table 14.1. Further, the study would illustrate how these values can contribute to sustainability (social, economic and environmental).

Thus, the values shown in the table also illustrates the variables endowed furthering sustainability to civil UAV deployment. Upholding all the values cited above are essential for sustainability (social, economic and environmental) which can further enhance responsible governance and deployment of the technology.

The next subsection on discussion and conclusion gives a brief illustration of the findings, how far the study has fulfilled the stated objectives and the research questions. The contribution of the paper to theory and practice is also highlighted.

14.7 Discussion and Conclusion

The study on civil UAV innovations in crop insurance application addresses issues and challenges associated with the governance of the technology. While addressing issues of responsibility and accountability in civil UAV governance, in-depth interviews were conducted based on two research questions – **How are the challenges of governance of civilian UAV innovations in crop insurance application can be addressed? How can the responsibility and accountability of the actors of innovation ensure effective governance?** In-depth interviews with the main actors of civil UAV innovations in crop insurance applications and extensive literature survey on the issues of governance, accountability and responsibility help in streamlining of gathered information and knowledge according to the objectives of the study.

However, for effective governance ensuring the culture-specific values of the society where the technology is deployed is an essential task. Around 69% of India population resides in villages, considering their values is crucial for responsible governance of civil UAV. Thus, the study found that upholding culture-specific values like transparency, trust, privacy, affordability, autonomy, effectiveness, safety and security promote sustainability (social, economic and environmental) further heralding responsible governance of the technology. The actors and institutions of governance of emerging technology promote innovations. The inefficiency of the actors and strict rules and regulations (institutions) at times also suppress or delays innovation. Effective collaboration and coordination among all the actors, slight flexibility of the institutions that regulate civil UAV deployment can promote trust, transparency and accountability furthering innovations of the technology in crop insurance applications.

Embedding values like responsibility and accountability make the actor of innovations answerable for his/her actions or undesirable consequences. The study is thus helping to see the issues of accountability, responsibility and governance of an emerging technology from the perspective of innovation framework. How the various actors and institutions of governance can promote or suppress innovations are also highlighted in the study. Thus, the paper is adding a new dimension to innovation studies juxtaposed by governance issues.

References

Ahrens J (2002) Governance and the implementation of technology policy in less developed countries. Econ Innov New Technol 11(4–5):441–476. https://doi.org/10.1080/10438590200000008

Aurora RS, Rachuri S (2016) Crop insurance. Int J Recent Res Asp 3(2):32–36

Axelsson R, Angelstam P, Degerman E, Teitelbaum S, Andersson K, Elbakidze M, Drotz MK (2013) Social and cultural sustainability: criteria, indicators, verifier variables for measurement and maps for visualization to support planning. Ambio 42(2):215–228. https://doi.org/10.1007/s13280-012-0376-0

Blyenburgh PV (1999) UAVs: an Overview. Air Space Eur I(5/6) 43–48

Bryson JM (2004) What to do when stakeholders matter: stakeholder identification and analysis techniques. Public Manag Rev 6(1):21–53. https://doi.org/10.1080/14719030410001675722
Devaney L (2016) Good governance? Perceptions of accountability, transparency and effectiveness in Irish food risk governance. Food Policy 62:1–10. https://doi.org/10.1016/j.foodpol.2016.04.003
DGCA (2016) DGCA-Civil_UAS (draft April 2016). Circular XX. Director General of Civil Aviation, New Delhi
DGCA (2018) Civil Aviation Requirements. Office Of the Director General of Civil Aviation, New Delhi
Dudovskiy J (n.d.) Snowball sampling. Research-Methodology (blog) Accessed 22 Oct 2018. https://research-methodology.net/sampling-in-primary-data-collection/snowball-sampling/
Finn RL, Wright D (2012) Unmanned aircraft systems: surveillance, ethics and privacy in civil applications. Comput Law Secur Rev 28(2):184–194. https://doi.org/10.1016/j.clsr.2012.01.005
Garg RG (2016) Insuring Indian farmers more effectively – ICT update. April 18, 2016. http://ictupdate.cta.int/2016/04/21/insuring-indian-farmers-more-effectively/
Giddens A (1999) Risk and responsibility. Mod Law Rev 62(1):1–10. https://doi.org/10.1111/1468-2230.00188
Gonzalez WJ (2015) New perspectives on technology, values, and ethics: theoretical and practical. Springer, New York
Guston DH (2012) The pumpkin or the Tiger? Michael Polanyi, Frederick Soddy, and anticipating emerging technologies. Minerva 50(3):363–379. https://doi.org/10.1007/s11024-012-9204-8
Hoven Van D, Vermaas J, Poel IV (2015) Handbook of ethics, values, and technological design – sources, theory, values and application domains|Jeroen van Den Hoven | Springer. 2015. https://www.springer.com/in/book/9789400769694
ICAO (2005) Global air traffic management operational concept. International Civil Aviation Organization. https://www.icao.int/Meetings/anconf12/Document%20Archive/9854_cons_en[1].pdf
Jepson P (2005) Governance and accountability of environmental NGOs. Environ Sci Pol 8(5):515–524. https://doi.org/10.1016/j.envsci.2005.06.006
Lee RG, Petts J (2013) Adaptive governance for responsible innovation. In: Responsible innovation. Wiley-Blackwell, pp 143–164. https://doi.org/10.1002/9781118551424.ch8
Lelea, Margareta Amy, Universität Kassel, and Wissenschaftliche Betriebseinheit Tropenzentrum (2014) Methodologies for stakeholder analysis for application in transdisciplinary research projects focusing on actors in food supply chains: reload reducing losses adding value. DITSL, Witzenhausen
Malerba F (2002) Sectoral systems of innovation and production. Res Policy 18:247–264
Mishra S, Singh R (2018) Responsible innovation: a new approach to address the theoretical gaps for innovating in emerging E-mobility sector. In: Ferri F, Dwyer N, Raicevich S, Grifoni P, Altiok H, Andersen HT, Laouris Y, Silvestri C (eds) Governance and sustainability of responsible research and innovation processes: cases and experiences, Springer briefs in research and innovation governance. Springer, Cham, pp 93–99. https://doi.org/10.1007/978-3-319-73105-6_12
Mobegi VO, Kaburi SN, Kombo A, Achuti DM, Matoke DA, Ombasa BB, Sewe Dr T (2012) Influence of new technology on financial performance; a case of small scale tea industry in Kebirigo, Kenya. 1(6):19
Mohamed N, Al-Jaroodi J, Jawhar I, Idries A, Mohammed F (2018) Unmanned aerial vehicles applications in future smart cities. Technol Forecast Soc Chang. https://doi.org/10.1016/j.techfore.2018.05.004
Nordmann A (2014) Responsible innovation, the art and craft of anticipation. J Res Innov 1(1):87–98. https://doi.org/10.1080/23299460.2014.882064
North DC (1991) Institutions. J Econ Perspect 5(1):97–112

Owen R (2013) In: Bessant JR, Heintz M (eds) Responsible innovation: managing the responsible emergence of science and innovation in society. John, Chichester
Parsai G (2015) Drones to help gauge crop damage – NATIONAL – The Hindu. October 6, 2015. https://www.thehindu.com/todays-paper/tp-national/drones-to-help-gauge-crop-damage/article7727876.ece
Pellizzoni L (2004) Responsibility and environmental governance. Environ Polit 13(3):541–565. https://doi.org/10.1080/0964401042000229034
Pennock JR (1952) Responsiveness, responsibility, and majority rule. Am Polit Sci Rev 46 (3):790–807. https://doi.org/10.2307/1952285
Peterson ME (2006) The UAV and the current and future regulatory construct for integration into the National Airspace System. J Air Law Commer 71:521–614
Rotolo D, Hicks D, Martin BR (2015) What is an emerging technology? Res Policy 44 (10):1827–1843. https://doi.org/10.1016/j.respol.2015.06.006
Salerno MS, de Vasconcelos Gomes LA, da Silva DO, Bagno RB, Freitas SLTU (2015) Innovation processes: which process for which project? Technovation 35:59–70. https://doi.org/10.1016/j.technovation.2014.07.012
Setiawan AD, Singh R (2015) Responsible innovation in practice: the adoption of solar PV in telecom towers in Indonesia. In: Koops B-J, Oosterlaken I, Romijn H, Swierstra T, van den Hoven J (eds) Responsible innovation 2. Springer, Cham, pp 225–243. https://doi.org/10.1007/978-3-319-17308-5_12
Siebert S, Teizer J (2013) Mobile 3D mapping for surveying earthwork using an unmanned aerial vehicle (UAV). In: Montreal, Canada. https://doi.org/10.22260/ISARC2013/0154
Sinha JP, Kushwaha HL, Kushwaha D, Singh N, Purushottam M (2016) Prospect of Unmanned Aerial Vehicle (UAV) Technology for Agricultural Production Management. In: IIT Kharagpur: Agricultural and Food Engineering Department, IIT Kharagpur
Stoker G (1998) Governance as theory: five propositions. Int Soc Sci J 50(155):17–28. https://doi.org/10.1111/1468-2451.00106
Swierstra T, Jelsma J (2006) Responsibility without moralism in technoscientific design practice. Sci Technol Hum Values 31(3):309–332. https://doi.org/10.1177/0162243905285844
Valavanis KP (ed) (2007) Advances in unmanned aerial vehicles: state of the art and the road to autonomy, International series on intelligent systems, control and automation: science and engineering, vol 33. Springer, Dordrecht
van Oudheusden M (2014) Where are the politics in responsible innovation? European governance, technology assessments, and beyond. J Res Innov 1(1):67–86. https://doi.org/10.1080/23299460.2014.882097
Von Schomberg R (2013) A vision of responsible research and innovation. In: Responsible innovation: managing the responsible emergence of science and innovation in society, pp 51–74
Xiang H, Tian L (2011) Development of a low-cost agricultural remote sensing system based on an Autonomous Unmanned Aerial Vehicle (UAV). Biosyst Eng 108(2):174–190. https://doi.org/10.1016/j.biosystemseng.2010.11.010
Zahinos A, Singh R, González-Benítez M (2013) Moving toward responsible innovation approach in the automotive industry: The SEAT Case 9

MIX
Papier aus verantwortungsvollen Quellen
Paper from responsible sources
FSC® C105338

If you have any concerns about our products,
you can contact us on
ProductSafety@springernature.com

In case Publisher is established outside the EU,
the EU authorized representative is:
Springer Nature Customer Service Center GmbH
Europaplatz 3, 69115 Heidelberg, Germany

Printed by Libri Plureos GmbH
in Hamburg, Germany